TECHNOCULTURE

CULTURAL ✌ POLITICS

A series from the Social Text collective

TECHNOCULTURE

Constance Penley and Andrew Ross, editors
(for the Social Text collective)

Cultural Politics, Volume 3

University of Minnesota Press
Minneapolis   Oxford

"Hacking Away at the Counterculture" first appeared in
Postmodern Culture, no. 1 (Fall 1989), © Andrew Ross. "Still Here"
reprinted from *Selected Poems of Langston Hughes,* © 1967 by
Langston Hughes. By permission of Alfred A. Knopf, Inc.

Published by the University of Minnesota Press
2037 University Avenue Southeast, Minneapolis, MN 55414
Printed in the United States of America on acid-free paper
Second printing, 1992

Library of Congress Cataloging-in-Publication Data

Technoculture / Constance Penley and Andrew Ross, editors.
    p.   cm. — (Cultural politics ; v. 3)
  Includes index.
  ISBN 0-8166-1930-1 — ISBN 0-8166-1932-8 (pbk.)
    1. Technology—Social aspects.   2. Communication and culture.
I. Penley, Constance, 1948-   .   II. Ross, Andrew.   III. Series:
Cultural politics (Minneapolis, Minn.) ; v. 3.
T14.5.T438   1991
303.48'3–dc20   90-25912
                                               CIP

A CIP catalog record for this book is available from the British
Library

Contents

Acknowledgments

Thanks are due to Jennifer Hayward and Ginger Strand, to the Princeton University Committee on Research in the Humanities and Social Sciences, and to the Susan B. Anthony Center for Women's Studies at the University of Rochester.

Introduction

At the very beginning of the student revolt, during the Free Speech Movement at Berkeley, the computer was a favorite target for aggression. . . . During the May events in Paris, the reversion to archaic forms of production was particularly characteristic. Instead of carrying out agitation among the workers with a modern offset press, the students printed their posters on the hand presses of the Ecole des Beaux Arts. The political slogans were hand-painted; stencils would certainly have made it possible to produce them en masse, *but it would have offended the creative imagination of the authors. The ability to make proper strategic use of the most advanced media was lacking. It was not the radio headquarters that were seized by the rebels, but the Odéon Theatre, steeped in tradition.*

Hans Magnus Enzensberger, 1971

I'd rather be a cyborg than a goddess.

Donna Haraway, 1984

Among the many surprises offered by the Chinese democracy movement of spring 1989 was the spectacle of the Western press cheerleading the students' use of new electronic media, especially fax machines, to communicate with students outside of Beijing. Most of the communications around the events, however, were one-way, composed primarily of the students' increasingly frantic gathering of information from Western media sources, as the suppression of the movement was stepped up. Since the hardware and the lines of communication were primarily in Western hands, very little information about the students' actual demands, desires, and strategies was *di-*

rectly relayed to the millions around the world who watched media coverage of the events in Tiananmen Square. For the most part, then, the students were receivers of Western information—more often than not, Western definitions of their own aims—rather than broadcasters of their own messages to allies and supporters in the West. The irony of this situation did not go entirely unnoticed. Much more prevalent, however, was the presentation of the students' use of the electronic media not only as a short-lived triumph for the "free flow of information" valorized by capitalist notions of free-market democracy, but also as transparent evidence of the Chinese people's *objective* desire for the kind of modernization that Western information technology represents. By contrast, there was no corresponding chorus of appreciation from the Western press when, during the *intifada*, Palestinians within the occupied territories made similar use of media technologies, including the fax machine, to communicate with the PLO; or when Saddam Hussein made strategic use of CNN's Baghdad broadcasts to frame his own communications with the world.

Notwithstanding the West's hosannas, the Chinese students were only picking up the legacy of what almost every liberation movement in the twentieth century has learned to do—turn the decentralizing and libertarian components of new media technologies against the state's centralized control of information. Vital to the Iranian Revolution (and to the subsequent Philippine ouster of President Marcos) was the widespread use of audiocassette technology to circulate speeches and teachings of the Ayatollah Khomeini. Radio technology was as crucial in the postwar struggles of African nation-states for independence as it was back in 1916 when the Easter Rising in Dublin centered on the post office, or when workers' and soldiers' councils in Berlin in 1919 took over military wireless networks to broadcast messages about their revolutionary activities.

The celebratory response to the student activities in China was clearly a consequence of Western control over media definitions of the events played out in Tiananmen Square. Similar conditions shaped coverage of antiapartheid activity in South Africa and, indeed, the Palestinian *intifada* in the occupied territories before Western media were all but banned from these countries. The struggle for self-determination, in each instance, was being waged under technological conditions produced elsewhere. A higher level of irony was tapped in the way in which technologies that hitherto had been widely used in China for purposes of cultural piracy—an anti-Western practice, in recent history—were now freshly presented as *samizdat* technologies, somehow now working in the service of Western ideals. It was a vivid illustration of the political efficiency with which "unscrupulous pirates" can easily become "freedom fighters" when it suits the Western powers.

Behind this miraculous redefinition lies a story about the uneven distribution of those new cultural technologies whose architectural capacity to copy, process, and stimulate makes nonsense out of the principles of individual property ownership of information. As a result of these technologies, the cultural piracy of Western literature, film and television images, popular music, and all kinds of information data flow has long been a standard practice in developing countries that lack the capital and the material apparatus for their own cultural production. From West Africa to the Philippines, the universal practice of piracy, whether state sanctioned or purely mercenary, has helped to deter the efforts of the transnational monopoly producers and brokers to control profitably and politically the flow of news, scientific data, educational materials, and entertainment. In some markets, such as the Indonesian music market, it is estimated that, out of the millions of cassettes and records of Western music sold each year, there is not one legitimate recording.[1] In many countries, attempts to enforce Western ideas about copyright and private intellectual property are seen as acts of cultural imperialism or, given the legacy of illiteracy accumulated over centuries of colonial rule, as postcolonial impertinence. The arrival of the technologies, then, is accompanied, as it were, by ideological instructions about their "proper" use that are often in direct contradiction to the obvious practical uses of the technologies. In most cases, the implied instructions, understood as deferring to Western notions of copyright, go unheeded.

In anticipation of a full-fledged New World Information and Communications Order (NWICO)—the step fostered by UNESCO toward informational sovereignty of all nation-states achieved through the multidirectional distribution of cultural and data flows—cultural piracy is often seen as an intermediate technology for developing nations. It is rarely what might properly be called an "appropriate technology," however, even if it contains and encourages elements of self-reliance and calls attention to local needs.[2] Just as any transfer of even small-scale nonappropriate technology from the West risks initiating a cycle of dependency upon Western parts and expertise, so too the target of cultural piracy has been almost exclusively Western culture itself, in the form of educational materials and information that are often hopelessly outdated because the supply flow of new products has been cut off by the Western source. At best, then, the piracy or hijacking of an inappropriate technoculture "from above" might help to stimulate attempts to build up local networks of cultural production and distribution. At worst, such practices merely reinforce cultural dependency upon Western ideas and opinions.

What is undeniable, however, is evidence of the wide array of strategies employed in developing countries to combat the monolithic picture of the "one-way flow" of Western technoculture that is often presented by the critics of transnational monopoly production. These strategies range from the official level of state- or publicly owned information services to the everyday level of popular *refunctioning* of foreign technologies or cultural products. Everybody knows one or two "cute" examples of this kind of refunctioning of technology: the Vietnamese farmers who turn bomb craters left by U.S. B-52 raids into fish ponds and rice paddies; the use of bicycles to run table saws, pump water, thresh rice and corn, and churn butter.[3] Just as important, but less apparent, is the complex process by which Western technoculture, even the most propagandistic and militaristic, is always being reread and reinterpreted in ways that make sense of local cultures and that intersect with local politics, with all sorts of results that go against the grain and the intentions of the Western producers and sponsors.

Public and political consciousness today is quite advanced about the "problems" and "solutions" experienced and generated in developing countries confronted by the domestic presence of transferred Western technocultures that tend, in Stephen Hill's phrase, "to burn like a cigarette on silken fabric."[4] Arguments about the erosion of national sovereignty and cultural identity, foreign dependency, and structural suppression of the right to communicate are commonly marshaled against Western attempts to enforce free-market ideology. As is sometimes the case, the critical left often spends more time debating and lamenting the effects of Western technoculture in other countries than it devotes to the conditions for creating technological countercultures in the West. Our own cultures, it is often assumed, have already been fully colonized by the cultural logic of technological rationality and domination, and so looking for signs of resistance is like looking for leftover meat in a lion's cage. Whatever is temporarily developed in the way of technologies of liberation is fated, or so the story goes, to be recuperated or reappropriated by the all-powerful sponsors of control technology.

One of the aims behind this volume is to resist that tendency of fatalistic thought, and to include other kinds of stories that do not automatically fall into line with the tradition of left cultural despair and alarmism. Consequently, the essays collected in *Technoculture* are almost exclusively focused on what could be called *actually existing technoculture* in Western society, where the new cultural technologies have penetrated deepest, and where the environments they have created seem almost second nature to us. Wary, on the one hand, of the disempowering habit of demonizing technology as a satanic mill of domination, and weary, on the other, of postmodernist celebrations of the technological sublime, we selected contributors whose critical knowledge might help to provide a realistic assessment of the politics—the dangers *and* the possibilities—that are currently at stake in those cultural practices touched by advanced technology.

We fully recognize that cultural technologies are far from neutral, and that they are the result of social processes and power relations. Like all technologies, they are ultimately developed in the interests of industrial and corporate profits and seldom in the name of greater community participation or creative autonomy. In many cases, the inbuilt principles of these technologies are precisely aimed at deskilling, information gathering, surveillance, and the social management of large populations. As a result, the research and development—mostly under military auspices—and the large-scale deployment of the new technologies tend to perpetuate capitalist modes of production and accumulation, the expropriation of cultural and technical skills, the international division of labor, social fragmentation, the policing of bodies, and the rationalization of nature. These perceptions have to be brought to bear against the picture of utopian social harmony promised by postindustrial ideologues, who preach that the new information and media technologies will bring an end to centralization and Fordist standardization, and will usher in participatory democracy based on interactive communications, electronic plebiscites, and culturally diverse communities, all achieved through the user-friendly agency of "clean" machines. The conflict between these two worldviews takes place in a climate where the horrors of environmental degradation are slowly forcing official policymakers to recognize, reluctantly, the need to set limits to technological growth.

Like most of the contributors to this volume, we, as editors, are conscientiously aligned with the technology-as-social-control school of thought and reject the postindustrialist fantasy of technical sweetness and light. Nonetheless, we recognize that the kinds of liberatory fantasies that surround new technologies are a powerful and persuasive means of social agency, and that their source to some extent lies in real popular needs and desires. Technoculture, as we conceive it, is located as much in the work of everyday fantasies and actions as at the level of corporate or military decision making. It is a mistake to dismiss such fantasies as false consciousness and such actions as compensatory bait, or to see their subjects as witless dupes of a smooth confidence trick. To deny the capacity of ordinary women and men to think of themselves as somehow in charge of even their most highly mediated environments is to cede any opportunity of making popular appeals for a more democratic kind of technoculture. More important is the task of re-creating a tradition of technoculture activism and practice that would be able to contest the pragmatic shape of these fantasies and everyday actions rather than dismiss them as the sugary fare of the lotus-eating masses. On the other hand, we consider it naive to think that the simply "nonpassive" use of a videocam, a VCR, a cassette recorder, or a personal computer constitutes a heroic act of resistance, or that it represents an achievement of political autonomy in itself. There may be little to be gained from simply adding to the paranoia and sense of victimization that is often produced by critics of the scary new panopticon, but there is arguably much more to be lost by asserting that the "leakiness" of panoptical systems proves that the sponsors of technological rationality are on the verge of being brought to their knees.

In short, *Technoculture* is presented in the knowledge that the odds are firmly stacked against the efforts of those committed to creating technological countercultures. But it is not always in our interests to lessen these odds further by theoretically reinforcing the conditions of helplessness and victimization. All the more reason, then, to unburden ourselves of some of the shopworn precepts of a cultural critique that depends, for its evidence, upon the kind of quantitative rationality that it presumably sets out to oppose. How many times, for example, have we heard it said, with rising contempt, that the average North American watches seven hours of television each day? Or that the citizens of developing countries are bombarded with a ceaseless flow of Western cultural goods? Or that government agencies have access to immense dossiers of information on average citizens? The cold facts may be scary, but they do not provide a basis for an alternative critical response because they pay lip service, finally, to a social and economic order that measures itself by the very rule of quantity and accumulation.

In fact, we know very little about the uses and purposes that modern television serves in people's lives; neither do we know what sense is being made of the actual programming that is attentively watched during the seven hours that TV sets may be switched on. So too, statistics about the quantity and volume of Western cultural products covering the globe tell us little about the interpretation of their messages on the part of wholly diverse non-Western populations. And, finally, we ought to consider that any centralized, "smart" supervision of an information-gathering system in the name of state surveillance would require, as Hans Magnus Enzensberger once argued, "a monitor that was bigger than the system itself": "a linked series of communications, or, to use the technical term, a switchable network, to the degree that it exceeds a certain critical size, can no longer be centrally controlled but only dealt with statistically." [5]

Statistically based critiques are inadequate, then, if they themselves appeal merely to the kind of rationality that is being questioned. So too, it is a mistake to conceive of technologies as hardware products alone, the engineers and designers of which have been hired or manipulated to implement the colonizing intentions of some powerful, conspiring group. If that were the case, then Orwell's seamlessly dystopian vision would have been realized long ago. Technologies are not repressively foisted upon passive populations, any more than the power to realize their repressive potential is in the hands of a conspiring few. They are developed at any one time and place in accord with a complex set of existing rules or rational procedures, institutional histories, technical possibilities, and, last, but not least, popular desires. All kinds of cultural negotiations are necessary to prepare the way for new technologies, many of which are not particularly useful or successful. It is the work of cultural critics, for the most part, to analyze that process, and to say how, when, and to what extent critical interventions in that process are not only possible but also desirable.

Let us remember that the story of modern technoculture is littered with honorable mentions: the early ham radio enthusiasts who helped to pioneer public wireless networks by thwarting military attempts to control communications; the hackers whose libertarian ethics and design skills helped to build the personal computer; the video artists and activists who used portapaks, initially developed for airborne reconnaissance in Vietnam, as effective instruments of countersurveillance; the pioneers of cable television who simply pirated signals, or the alternative media collectives that rent satellite transponders; independent, radical desktop publishers; audio and video scratchers, mixers, samplers, and appropriators; fanzine producers; community radio stations; CB culture; bulletin board systems operators; Peacenet; *samizdat* technologists; revolutionary sources like Radio Venceremos; all of those who use the technologies of instant printing, photocopying, cassette taping, radio transmission, and video production, which are as crucial to local, decentralized, community activism as they are instrumental to the task of building up networks of national and international resistance. To take only one example, anyone who has participated in actions organized by the AIDS activist group ACT UP, one of the exemplary models of activism in our time in its fight against the discriminatory path of official "science and technology" through the health care system, will have seen how indispensable these technologies are for the purposes of countersurveillance, distribution of information and iconic messages, manipulative use of mainstream television and journalism, and safe sex and health education. Activism today is no longer a case of putting bodies on the line; increasingly, it requires and involves bodies-with-cameras.

Some of the examples mentioned above are well-known stories about politically articulate uses of technology, and many of them are either examined or alluded to by contributors to this volume. But we have also found it important to include the category of *protopolitical* technoculture, which covers the complex psychosocial process by which people, either individually or in groups, make their own independent sense of the stories that are told within and about an advanced technological society. Unorganized and, more often than not, politically mute, these less heroic, daily activities are an important populist substratum of the political process by which we might seek to turn technocommodities into resources for waging a communications revolution from below.

Consequently, the groups or cultures that are examined here cover a very broad spectrum: high-tech office workers, *Star Trek* fans, Japanese technoporn producers, teenage hackers, AIDS activists, rap groups, pregnant women, video and media activists, political artists, rock stars, and science fiction writers. The theorists, activists, artists, and scholars who have contributed to *Technoculture* are also drawn from a wide range of disciplines and professions. Some of them are concerned with the production or management of repressive technocultures, some draw attention to the politics of creative appropriation, and others consider both sides of the question. All seem to be committed, however, to the pressing need for more, rather than less, technoliteracy—a crucial requirement not just for purposes of postmodern survival but also for the task of decolonizing, demonopolizing, and democratizing social communication.

All of the contributors, of course, speak for themselves, and as with all selective collections of essays, restrained by time, space, and limited editorial clout on our part, there are many areas of the field of technoculture that are not represented or covered, and that warrant many pages of analysis in their own right. The largest single area of concern is ecology, where public consciousness about the theory and practice of alternative technologies has solidified the demand for environmental responsibility in every sphere of life. It is there that the contradictions between the ecosphere and the technosphere are increasingly exposed in everyday culture and, in many respects, constitute our immediate political horizon in the 1990s. Other, more specific topics would include the highly developed technoculture of the handicapped and the complexity of their discussions around appropriate levels of technology, and the diverse relations to technology that are negotiated by the elderly, or by children. While these areas, and others, fall outside the range of this particular volume, we feel that they are nonetheless an important part of the debate that the book addresses.

NOTES

1. John Chesterman and Andy Lipman, *The Electronic Pirates: DIY Crime of the Century* (London: Routledge, 1988), 43.

2. George Ovitt, Jr., "Appropriate Technology: Development and Social Change," *Monthly Review,* 40, 9 (February 1989).

3. Ibid., 30.

4. See Stephen Hill, *The Tragedy of Technology: Human Liberation versus Domination in the Late Twentieth Century* (London: Pluto, 1988).

5. Hans Magnus Enzensberger, "Constituents of a Theory of the Media," *The Consciousness Industry,* trans. Stuart Hood (New York: Seabury, 1974).

Cyborgs at Large:
Interview with Donna Haraway
Constance Penley and Andrew Ross

Andrew Ross: Many people from different audiences and disciplines came to your work through "A Manifesto for Cyborgs," which has become a cult text since its appearance in *Socialist Review* in 1985. For those readers, who include ourselves, the recent publication of *Primate Visions* and the forthcoming *Simians, Cyborgs, and Women* provides the opportunity to see how your work as a historian of science was always more or less directly concerned with many of the questions about nature, culture, and technology that you gave an especially inspirational spin to in the Cyborg Manifesto. So we'd like to begin with a more general discussion of your radical critiques of the institutions of science. Although you often now speak of having been a historian of science, almost in the past tense, as it were, it's also clear that you have many more than vestigial loyalties to the goals of scientific rationality — among which being the need, as you put it, in a phrase that goes out of its way to flirt with empiricism, the need for a "no-nonsense commitment" to faithful accounts of reality. Surely there is more involved here than a lingering devotion to the ideals of your professional training?

Donna Haraway: You've got your finger right on the heart of the anxiety — some of the anxiety and some of the pleasure in the kind of political writing that I'm trying to do. It seems to me that the practices of the sciences — the sciences as cultural production — force one to accept two simultaneous, apparently incompatible truths. One is the historical contingency of what counts as nature for us: the thoroughgoing artifactuality of a scientific object of knowledge, that which makes it inescapably and radically contingent. You peel away all the layers of the onion and there's nothing in the center.

And simultaneously, scientific discourses, without ever ceasing to be radically and historically specific, do still make claims on you, ethically, physically. The objects of these discourses, the discourses themselves, have a kind of materiality; they have a sort of reality to them that is inescapable. No scientific account escapes being story-laden, but it is equally true that stories are not all equal here. Radical relativism just won't do as a way of finding your way across and through these terrains. There are political consequences to scientific accounts of the world, and I remain, in some ways, an old-fashioned Russian nihilist. My heroes are the women who set off to get agronomy and medical degrees in Zurich in the 1860s, and then went back to serve the revolutionary moment in Russia with their scientific skills. A lot of my heart lies in old-fashioned science for the people, and thus in the belief that these Enlightenment modes of knowledge *have* been radically liberating; that they give accounts of the world that can check arbitrary power; that these accounts of the world ought to be in the service of checking the arbitrary. I hold onto that simultaneously with an understanding that I learned from the discipline of the history of science, that the sciences are radically contingent. They are specific historical and cultural productions. So, I felt like a political actor and scholar trying to hold those two things together when the disciplines, as well as social movements, want to pull them apart.

AR: So there remains a sense of responsibility to provide reliable knowledge about the world. In your new book, you push this responsibility to what might appear, to some, to be rather bizarre lengths. At one point, you say that to have a better account of the world—the laudable goal of science after all . . .

DH: Yes, which one always says with a nervous laugh . . .

AR: To have that better account of the world, you propose that we ought to be able to see the world and its objects as agents to the extent that we ought to be aware of what you call "the world's independent sense of humor." We're curious about this phrase, and

were wondering if you could give us a more concrete example of the world's "sense of humor."

DH: Someone asked if I meant the earthquake. [laughter; this interview took place in Santa Cruz, California, soon after the 1989 San Francisco Bay Area earthquake] Well, obviously what's going on there is some kind of play with metaphors. In this respect, I'm most influenced by Bruno Latour's actor-network theory, which argues that in a sociological account of science all sorts of things are actors, only some of which are human language-bearing actors, and that you have to include, as sociological actors, all kinds of heterogeneous entities. I'm aware that it's a risky business, but this imperative helps to break down the notion that only the language-bearing actors have a kind of agency. Perhaps only those organized by language are *subjects,* but agents are more heterogeneous. Not all the actors have language. And so that presents a contradiction in terms, because our notions of agency, action, and subjectivity are all about language. So you're faced with the contradictory project of finding the metaphors that allow you to imagine a knowledge situation that does not set up an active/passive split, an Aristotelian split of the world as the ground for the construction of the agent, or an essentially Platonist resolution of that, through one or another essentialist move. One has to look for a system of figures to describe an encounter in knowledge that refuses the active/passive binary which is overwhelmingly the discursive tradition that Western folks have inherited. So you go for metaphors like the coyote, or trickster figure. You go for odd pronouns, which encourage an acknowledgment that the relationship between nature and human is a social relationship for which none of the extant pronouns will do. Nature in relation to us is neither "he," "she," "it," "they," "we," "you," "thou" . . . and it's certainly not "it." So you're involved in a kind of science fictional move, of imagining possible worlds. It's always important to keep the tension of the fiction foregrounded, so that you don't end up making a kind of animist or pantheist claim. There's also the problem, of course, of having inherited a particular set of descriptive technologies as a Eurocentric and Euro-American person. How do I then act the *bricoleur* that we've all learned to be in various ways, without being a colonizer; picking up a trickster figure, for example, out of Native American stories? How do you avoid the cultural imperialism, or the orientalizing move of sidestepping your own descriptive technologies and bringing in something to solve your problems? How do you keep foregrounded the ironic and iffy things you're doing and still do them seriously? Folks get mad because you can't be pinned down, folks get mad at me for not finally saying what the bottom line is on these things: they say, well do you or don't you believe that nonhuman actors are in some sense social agents? One reply that makes sense to me is, the subjects are cyborg, nature is coyote, and the geography is elsewhere.

AR: It seems that you are increasingly, in your work, sympathetic to the textualist or constructionist positions, but it's clear also that you reject the very easy path of radical constructionism, which sees all scientific claims about the object world as merely persuasive rhetoric, either weak or strong depending on their institutional success in claiming legitimacy for themselves. Your view seems to be: that way lies madness . . .

DH: Or that way lies cynicism, or that way lies the impossibility of politics. That's what worries me.

AR: And your way of retaining political sanity is?

DH: Politics rests on the possibility of a shared world. Flat out. Politics rests on the possibility of being accountable to each other, in some nonvoluntaristic "I feel like it today" way. It rests on some sense of the way that you come into the historical world encrusted with barnacles. Metaphorically speaking, I imagine a historical person as being somehow like a hermit crab that's encrusted with barnacles. And I see myself and everybody else as sort of switching shells as we grow. [laughter] But every shell we pick up has its histories, and you certainly don't choose those histories—this is Marx's point about making history, but not any way you choose. You have to account for the encrustations and the inertias, just as you have to remain accountable to each other through learning how to remember, if you will, which barnacles you're carrying. To me, that is a fairly straightforward way of avoiding cynical relativism while still holding on, again, to contingency.

Constance Penley: In an essay on the history of the sex/gender split you argue that one of the unfortunate results of the antiessentialist position of feminist constructionists is that biology (which you equate with the "sex" side of the sex/gender split) has been un-

dervalued as a realm of investigation, where it really ought to have been seen as a much more active site for contesting definitions of "nature" that concern women quite directly. We can see where a sustained investigation of biology is useful for revealing historical and ideological links within science between "nature" and "femininity," but we'd like you to say what role you see biology playing in the future "reinvention of nature."

DH: This is actually very close to my heart, because there's that cryptobiologist lurking under the culture critic. The simplest way to approach that question is by remembering that biology is not the body itself, but a discourse. When you say that my biology is such-and-such—or, I am a biological female and so therefore I have the following physiological structure—it sounds like you're talking about the thing itself. But, if we are committed to remembering that biology is a *logos,* is literally a gathering into knowledge, we are not fooled into giving up the contestation for the discourse. I subscribe to the claim of Foucault and others that biopolitical modes of fields of power are those which determine what *counts* in public life, what counts as a citizen, and so on. We cannot escape the salience of the biological discourses for determining life chances in the world—who's going to live and die, things like that, who's going to be a citizen and who's not. So not only do we literally have to contest for the biological discourses, there's also tremendous pleasure in doing that, and to do that you've got to understand how those discourses are enabled and constrained, what their modes of practice are. We've got to learn how to make alliances with people who practice in those terrains, and not play reductive moves with each other. We can't afford the versions of the "one-dimensional-man" critique of technological rationality, which is to say, we can't turn scientific discourses into the Other, and make them into the enemy, while still contesting what nature will be for us. We have to engage in those terms of practice, and resist the temptation to remain pure. You do that as a finite person, who can't practice biology without assuming responsibility for encrusted barnacles, such as the centrality of biology to the construction of the raced and sexed bodies. You've got to contest for the discourse from within, building connections to other constituencies. This is a collective process, and we can't do it solely as critics from the outside. Gayatri Spivak's image of a shuttle, moving between inside and outside, dislocating each term in order to open up new possibilities, is helpful.

CP: Well, this brings us to the role of the Cyborg Manifesto in the "reinvention of nature." One of the most striking effects of the Cyborg Manifesto was to announce the bankruptcy of an idea of nature as resistant to the patriarchal capitalism that had governed the Euro-American radical feminist counterculture from the early 70s to the mid-80s. In the technologically mediated everyday life of late capitalism, you were pointing out that nature was not immune to the contagions of technology, that technology was part of nature conceived as everyday social relations, and that women, especially, had better start using technologies before technology starts using them. In other words, we need techno-realism to replace a phobic naturalism. Do you see the cyborg formulation of the nature/technology question as different from, or falling into the same alignment as, the nature/culture question that you had spent much more of your time exploring as a historian of science?

DH: That's an interesting way to put it. I'm not sure what to say about that. What I was trying to do in the cyborg piece, in the regions that you're citing there, is locate myself and us in the belly of the monster, in a technostrategic discourse within a heavily militarized technology. Technology has determined what counts as our own bodies in crucial ways—for example, the way molecular biology had developed. According to the Human Genome Project, for example, we become a particular kind of text which can be reduced to code fragments banked in transnational data storage systems and redistributed in all sorts of ways that fundamentally affect reproduction and labor and life chances and so on. At an extremely deep level, nature for us has been reconstructed in the belly of a heavily militarized, communications-system-based technoscience in its late capitalist and imperialist forms. How can one imagine contesting for nature from that position? Is there anything other than a despairing location? And, in some perverse sense which, I think, comes from the masochism I learned as a Catholic, there's always the desire to want to work from the most dangerous place, to not locate oneself outside but inside the belly of the monster. [laughter]

It's not that I think folks who are doing other kinds of work more directly oppositional, more critical of technological discourse, aren't doing important work; I think they often are. But I want myself and lots of other people to be inside the belly of the monster, trying to figure out what forms of contestation for nature can exist there. I think that's different from reproducing the cultural appropriation of nature, reducing nature yet again to a source redefined culturally. Without the nature/culture split, how can nature be reinvented, how can you make those moves? In my more recent work, for example, on the discourse of immune systems, that means discovering extraordinarily rich resources for avoiding the narrative of the invaded self, the defended, walled city invaded by the infecting Other. These discourses have the potential for telling very different stories about relationality, connection and disconnection in the world. We need to ask how those kinds of extant languages, practices, resources in immunology could become more determinative of the practices of medicine.

AR: **We'd like to try to clear up some of the more obvious misreadings that no doubt have been attached to your notion of the cyborg.**

DH: [laughs] Yes, let's.

AR: **It seems clear that there are good cyborgs and there are bad cyborgs, and that the cyborg itself is a contested location. The cyborgs dreamed up by the artificial intelligence boys, for example, tend to be technofascist celebrations of invulnerability, whereas your feminist cyborgs seem to be more semipermeable constructions, hybrid, almost makeshift attempts at counterrationality. How do you prevent, or how do you think about ways of preventing, cyborgism from being a myth that can swing both ways, especially when the picture of cyborg social relations that you present is so fractured and volatile and bereft of secure guarantees?**

DH: Well, I guess I just think it *is* bereft of secure guarantees. And to some degree, it's a refusal to give away the game, even though we're not entering it on unequal terms. It is entirely possible, even likely, that people who want to make cyborg social realities and images to be more contested places—where people have different kinds of say about the shape of their lives—will lose, and are losing all over the world. One would be a fool, I think, to ignore that. However, that doesn't mean we have to give away the game, cash in our chips, and go home. I think that those are the places where we need to keep contesting. It's like refusing to give away the notion of democracy to the right wing in the United States. It's like refusing to leave in the hands of hostile social formations tools that we need for reinventing our own lives. So I'm not, in fact, all that sanguine. But, (a) I don't think I have a lot of choice, and I know we lose if we give up. And (b) I know that there's a lot going on in technoscience discourses and practices that's not about the devil, that's a source of remarkable pleasure, that promises interesting kinds of human relationships, not just contestatory, not always oppositional, but something often more creative and playful and positive than that. And I want myself and others to learn how to describe those possibilities. And (c) even technoscience worlds are full of resources for contesting inequality and arbitrary authority.

CP: Your image of the cyborg paradoxically both describes what you see as a new, actually existing, hybrid subjectivity and offers a polemical, utopian vision of what that new subjectivity ought to be or will be. In other words it's something actually existing now but also an image . . .

DH: A possible world.

CP: A possible world. But our question is really not about the paradox, because we think the paradox is a suggestive and productive one. Most utopian schemes hover somewhere in between the present and the future, attempting to figure the future as the present, the present as the future. Rather, we're wondering if the way you have constructed your cyborg leaves any room for anything that could be called "subjectivity," and what the consequences of that possible omission may be. In other words, how useful to us now—"us" meaning socialist-feminists—is a myth or model that asks us to think and theorize without the categories of sexual difference, infantile sexuality, repression, and even the unconscious, because it is clear that your cyborg wants to have no truck with anything as nineteenth-century and archaic as the unconscious?

DH: Well, I think that might have been true in 1985; I was more of a fundamentalist about psychoanalysis than I am now, partly having been worn down by all my psychoanalytic buddies. [laughs] But my resistance to psychoanalysis is very much like my resistance to the church. I really think I've been vaccinated. Precisely because of understanding the power of a truly totalizing dogma that can include all stories, and my sense that the psychoanalytic narratives as they have been developed in the human sciences and in feminism, have a potential that I recognize with my vaccinated soul . . .

CP: **When I read *Primate Visions* I have to say that it really gave me a much stronger sense of why it was so important for you to come up with a creature that wasn't about Oedipal subjectivity . . .**

DH: Yes, which isn't quite the same thing as coming up with a creature without an unconscious. As a strategic and emotional matter, I really am hostile to the Oedipal accounts and their mutants—not because I don't recognize their power but because I am *too convinced* of their power. Again, it's the problem of being in the belly of the monster and looking for another story to tell, say, about some kind of creature with an unconscious that can nonetheless produce the unexpected, that can trip you, or trick you. Can you come up with an unconscious that escapes the familial narratives; or that exceeds the familial narratives; or that poses the familial narratives as local stories, while recognizing that there are other histories to be told about the structuring of the unconscious, both on personal and collective levels. The figures that we've used to structure our accounts of the unconscious so far are much too conservative, much too heterosexist, much too familial, much too exclusive. Much too restricted, also, to a particular moment in the acquisition of language; I think there are many kinds of acquisition of language throughout life, coming into history in different ways that isn't the same thing as coming into the familial. This all sounds very utopian, but I end up wanting a psychoanalytic practice—which I don't do myself—that recognizes the very local and partial quality of the Oedipal stories. Instead I see them cannibalizing too much of what counts as theoretical discourse. They're very powerful cannibalizers because they're very good stories. And I know in my heart that by analogy, I could have remained a Roman Catholic and thought anything I wanted to think if I was willing to put enough work into it, because these universal stories have that capacity, they really can accommodate anything at all. At a certain point you ask if there isn't another set of stories you need to tell, another account of an unconscious. One that does a better job accounting for the subjects of history. It's true that the '85 cyborg is a little flat, she doesn't have much of an unconscious.

CP: Well, it doesn't have the unconscious of the Oedipal stories because you've removed that. But, perhaps too it doesn't have that which in the unconscious resists . . .

DH: And that's a bigger problem . . .

CP: precisely the imposition of those Oedipal narratives.

DH: In some ways, I tried to address this in my notion of "situated knowledge," which, with the coyote, brings in another set of story cycles, where there is a resistance and a trickster, producing the opposite of—or something other than—what you thought you meant. Some kind of operator that tricks you, which is what I suppose the unconscious does . . .

CP: Maybe a trickster cyborg!

DH: Something like that.

CP: Along the same lines, we were especially wondering if your wish to construct a "myth" or model that makes an end run around Oedipal subjectivity and the unconscious is in fact the best one for ensuring that socialist-feminism take into account the mechanisms of racism—which is one of the most important aspects of your project. You look to the fiction of black science fiction writer Octavia Butler to give us "some other order of difference . . . that could never be born in the Oedipal family narrative." This new order of difference—and these are your words—is "about miscegenation, not reproduction of the One," because Butler's characters interbreed and create new gene pools across not only races but species. In other words, cyborg subjectivity will be hybridized, mixed and plural . . .

DH: What you never have with Butler is the original story. You never have the primal scene. You always have the chimeric . . .

CP: Right. So you end up with a subjectivity that's hybridized, mixed, and plural, rather than split.

DH: That's exactly right.

CP: But doesn't something get lost in our understanding of the dynamics of racism when we eliminate the split subject? If we no longer have a subject of the unconscious, this makes it difficult if not impossible to give an account of psychical mechanisms like displacement, projection, fetishism, which writers like Frantz Fanon or Homi Bhabha would consider crucial terms for being able to explain the dynamic of the psychic structure of racism.

DH: I believe it is correct that you can't work without a conception of splitting and deferring and substituting. But I'm suspicious of the fact that in our accounts of both race and sex, each has to proceed one at a time, using a similar technology to do it. The tremendous power and depth of feminist theories of gender in the last ten or fifteen years could not have been achieved without psychoanalysis. Similarly, I think you're right that Bhabha and Fanon and some others could not have worked without those tools in understanding race. But it has remained true that there is no compelling account of race and sex *at the same time*. There is no account of any set of differences that work other than by twos simultaneously. Our images of splitting are too impoverished. Consequently, we say, almost ritualistically, things like "We need to understand the structuring of race, sex, class, sexuality, et cetera." While these issues are related to one another, we don't actually have the analytical technologies for making the connections.

So, when I draw from a writer like Octavia Butler, or a theorist like Hortense Spillers, I try to say the following. Those people who have, in fiction and in theory, laid out for us the conditions of captivity in slavery in the New World have among other things done something very important to our theories of psychoanalysis. They have said (and here I'm borrowing primarily from Spillers, who is saying the same thing lots of African-Americans have said for years) that the situation of the human being in slavery is the situation of the body that passes on the status of "nonhuman" to the children; it is the story of the people who exist outside the narratives of kinship. The white woman married the white man; he had rights in her that she didn't have in herself. She was a vehicle for the transmission of legitimacy, so she was precisely the vehicle for the transmission of the Law of the Father. The person in captivity, however, did not even enjoy the status of being human. The mother passed on her status, not her name, to the child, not the father; and the status of the mother was not human. And it is precisely that historical and discursive situation which, in Spiller's language, positions black men and women outside the system of gender governed by the Oedipal story of incest and kinship.

Those are the people—the hybrid peoples, the conquest peoples, the enslaved peoples, the nonoriginal peoples, and the dispossessed Native Americans—who populated and *made* the New World. If you retell the history of what it means to be white, then you see the perversion of the compulsion to reproduce the sacred image of the Same: the compulsion of race purity and the control of women for the reproduction of race purity. And if you foreground the stories of captivity and conquest and nonoriginality, the New World then has a different set of stories attached to it. Now I think that these are stories that very much involve an unconscious structuring, that they are unconscious structurings that really do throw into question the relationships of gender and race.

Octavia Butler is a very frustrating writer in some ways, because she constantly reproduces heterosexuality even in her polygendered species. But I am drawn to the "nonoriginality" of her characters: as diasporic people, they can't go back to an original that never existed for them, and therefore they are not embedded in the system of kinship as theorized by Freud and Lévi-Strauss. Too much of Anglo feminist theory has started out from Freud, Lévi-Strauss, and Lacan. And I think that's unfortunate.

AR: On that note, we'd like to question you on the rhetorical force of the phrase "We are all cyborgs." On the one hand, it seems to be a general description of women's situation in the advanced technological conditions of postmodern life in the First World. On the other hand, it seems to function like the kind of identificatory statement or gesture which is often made in support of oppressed or persecuted groups, like "We are all Jews," or, now, "We are all Palestinians." It's difficult not to think of this latter sense in terms of the specifically Asian women of color whose labor primarily is the basis of the microelectronics revolution, and who, in your essay, seem to be privileged as cyborgs that are somehow more "real," say, than First World feminist intellectuals.

DH: Which, I agree, won't work. My narrative partly ends up further imperializing, say, the Malaysian factory worker. If I were rewriting those sections of the Cyborg Manifesto I'd be much more careful about describing who counts as a "we," in the statement, "We are all cyborgs." I would also be much more careful to point out that those are subject positions for people in certain regions of transnational systems of production that do not easily figure the situations of other people in the system. I was using Aihwa Ong's work there, in her remarkable study, in *Spirits of Resistance,* since published, of Malaysian factory workers in the Japanese technoscience-based multinationals. A U.S. immigrant, Ong was born an ethnic Chinese woman in Malaysia and was adopted by a Malay family when she did her ethnographic fieldwork for her Ph.D. from Columbia University. She writes about young women whose families acquired the colonial status of "Malay" when the British imported Javanese immigrants to create a Malay peasant yeomanry for subsistence food production in the plantation economy of British Malaysia. Consequently, to be native Malay was already to be the product of a colonial migration, subsequently repositioned in the Malaysian state in the 1970s in ethnic contests, among other things, between the Malay and the Chinese. At that time, a whole nationalist discourse foregrounded the ethnic status of Malay, and promoted the look-East policy to Japanese transnationals rather than to American transnationals. What kind of personal and historical subjectivity did the young women in these factories develop? This is an incredibly contradictory situation, and naming those contradictions seems to be crucial now; to name them "cyborg" seems to me more iffy. I think what I would want is more of a family of displaced figures, of which the cyborg is one, and then to ask how the cyborg makes connections with these other nonoriginal people (cyborgs *are* nonoriginal people) who are multiply displaced. Could there be a family of figures who would populate our imagination of these postcolonial, postmodern worlds that would not be quite as imperializing in terms of a single figuration of identity?

AR: **Would your talking about that family, today, also be posed less in terms of the rhetoric of survivalism? This critique is often made of intellectuals who speak about working-class people, especially from another culture, as if their situation were primarily one of survivalism . . .**

DH: You mean, "Well gee thanks, but we live a more fully human life than that." I think that's right, the survivalist rhetoric doesn't give enough space to more than survival, to the way people live complicated lives that aren't simply about insertion in your system of explanation.

AR: **This is a kind of rhetoric and style question; it's about the manifesto format itself, which has its own particular "generic" demands historically. Your manifesto, as we read it, is about as far from** being programmatic, in the sense of espousing a party line, as could be. But it's also more legislatist than the manifestos of, say, artistic avant-garde movements. There's a kind of poetic license there, on the one hand, that is a curious bedfellow to the otherwise earthy sense of political realism, which is on the other side. What is your experience of finding readers who might have been bewildered by that heady mix?

DH: Yeah, I've been surprised by the reactions of readers, because I had almost no idea whether people would read this. The original assignment was to do five pages on what socialist-feminist priorities ought to be in the Reagan years. The *Socialist Review* collective asked a whole lot of people identified as socialist-feminist. The writing of that piece immersed me for a whole summer in the process of finding a set of metaphors I didn't know were there. It was a summer about writing. I didn't set out to write a manifesto, or to write what turned out to be a heavily poetic and almost dream-state piece in places. But, in many ways, it turned out to be about language. As a result, the manifesto is not politically programmatic in the sense of proposing a priority of options; it's more about all kinds of linguistic possibilities for politics that I think we (or I) haven't been paying enough attention to.

CP: Yes . . . that comes across. We'd like to focus now on popular practices, rather than intellectual debates about women and technology. One of the intended aims of your work is helping women to overcome their culturally induced technophobia. You do so through getting readers excited about specific areas of science that have heavily involved women, like primatology; by frequently citing utopian science fiction narratives by women like Joanna Russ and Octavia Butler that offer empowering visions of a new relation to gender, race, nature, and technology; and by imaginatively demonstrating, in the Cyborg Manifesto, that we are already cyborgs, already creatures that are wondrously both human and technological. Do you see any evidence in everyday life that women are in fact overcoming their technophobia? What seem to be the present possibilities and difficulties?

DH: That's tough, because if you go at it statistically—look at the most recent National Science Foundation statistics, for example—fewer women are getting engineering degrees, fewer women are entering science programs than was true ten and fifteen years ago. The gains of women, in the sixties and seventies, as practitioners of science and engineering have been eroded, and the same thing is true for people of color, men and women alike. There has been a massive retrenchment in affirmative action programs, especially for loans for students at entering stages; heavier pressures on families; all exacerbated by the economics of Reaganism and Thatcherism. While there are a few little gains here and there, overall we're losing again. On the other hand, I also see, for example, in the development of ecofeminism some very savvy new relations to science and technology developing at the level of popular practices.

AR: How about everyday practices in the household, or in non-professional spaces?

DH: I think I would like to know a lot more about them. What we need are thick ethnographic accounts of those very practices, in various social, regional, ethnic, racial settings. I could tell anecdotes of women I know who have achieved wonderfully heterogeneous kinds of technological literacy, but I have no idea what that means in terms of broader social issues.

AR: In general, you're very much opposed to any kind of holistic response to the tyrannies of technological rationality. How do you expect a philosophy of partialism—which at least is one of the ways of describing your position—to become a popular philosophy? Especially in an age in which millions upon millions of people have been attracted by the holistic principles of New Age movements and practices, from the practice of alternative therapies to the intense fascination with the scientist-cum-mystic who meditates about quantum physics. That sort of holism is not exactly the kind of antiscience metaphysics that your cyborgism condemns . . .

DH: No, it's a kind of mirror image.

AR: Right. A mythology of alternative science that is deeply in love with science. But if the appeal of holism runs so deep for people who want to resolve a sense of loss or absence in their lives, how can cyborgism make headway in contesting that kind of popular appeal of science's promise of completion for people?

DH: That's a tough one. It might come down to this. How can there be a popular, playful, and serious imaginative relation to techno-science that propounds human *limits* and *dislocations*—the fact that we die, rather than Faustian (and so deadly)—evasions. Again, this might be a psychoanalytic question, since those holisms have the appeal of bridging all the parts and promising an ultimate oneness. They promise what they cannot, of course, deliver, or only pretend to deliver at the cost of deathly practices, almost a worship of death. The kind of partiality I'm talking about is resolutely antitranscendentalist and antimonotheist, fully committed to the fact that we don't live after we die. In religious language, that's what it comes down to: no life after death. Any transcendentalist move is deadly; it produces death, through the fear of it. These holistic, transcendentalist moves promise a way out of history, a way of participating in the god trick. A way of denying mortality.

On the other hand, in the face of having lived forty-five years inside nuclear culture, in the face of the kind of whole-earth threat issuing from so many quarters, it's clear that there is a historical crisis of the sort that might really be able to shake the hold of these monotheisms. Some deep, inescapable sense of the fragility of the lives that we're leading—that we really do die, that we really do wound each other, that the earth really is finite, that there aren't any other planets out there that we know of that we can live on, that escape velocity is a deadly fantasy. What's also clear from popular culture is that large numbers of people are at least aware of the crisis we're facing, a crisis of historical consciousness where the master narratives will no longer soothe as they have for a couple thousand years, in Christian culture at any rate.

AR: As cultural critics we often find that the kind of vanguardist culture criticism that tends to focus on vanguardist texts can very easily embrace partialism and a philosophy of subjectivity that doesn't depend upon secure identifications. But when you deal with popular practices . . .

DH: You're in a different world . . .

AR: it's usually the opposite: the circuit of identification tends to be more important, and probably necessary to the affective appeal of the text.

DH: The whole technology of pleasure works that way. I recognize what you're saying, but there's also a part of me that's a little bit unsure about the generalization. I can think of someone like Ursula Le Guin, who's a very, very popular writer. And you can read Ursula Le Guin either as a holist who has a sense of an earth that can remain unviolated — a kind of naturalist holism — but she equally cries out to be read in terms of her insistence on limitation. It's not holism she's insisting on, but rather on fragility and limitation by avoidingnarratives of completion. The pleasure of her stories derives from being reminded of one's materiality — the pleasure of being at home in the world, rather than needing transcendence from it. And being at home in the world is about a kind of partiality: you just plain aren't everywhere at once. To relocate, you have to dislocate.

[AR's pained laughter]

CP: **There is the same kind of split in more popular practices — a wish for holism and completion, but at the same time an incredible play with the idea of partiality.**

AR: **And a sense of relief?**

DH: Relief, and sanity, that you can let go of an illusion which had felt necessary. The mother of my lover is a person who had been interested in channelers and various New Age phenomena. I don't think it's very useful to think of her as someone who needs some kind of scientist transcendentalism: I don't think those are her pleasures, which are more like science fictional pleasures — of imagined historical connections into pasts and futures — not at all about being masterful, or being in charge of the whole.

CP: **What we especially like in the Cyborg Manifesto is the use of the term *scary* to describe the new informatics of domination that sponsors of advanced technology have installed everywhere. It suggests a nightmarish quality, but it also hints at excitement and adventure, especially girls' adventures in realms hitherto off-limits to them. In this respect, it seems to be different from the note of technoparanoia usually sounded in orthodox left accounts of tech surveillance and social control. There's a fictional action-adventure cast to your version of "scariness" . . .**

DH: The funhouse!

CP: **that more accurately reflects the everyday response of ordinary people to control technologies, rather than the paranoid vi-**

sion of unrelieved domination everywhere. If you agree with that characterization of scariness, does that mean to say that you don't think we ought to be too scared?

DH: Certainly not fear unto death. Paranoia bores me. It's a psychopathology, and it's an incredibly indulgent one. It sees the Eye everywhere, and it strikes me as a kind of arrogance. The paranoid person takes up too much social space, their friends have to take care of them all the time, and it's a lousy model for how we ought to be feeling collectively . . . so I agree about the rejection of paranoia in the face of the panopticon of postmodernism, or the "polyopticon," or whatever you want to call it. The funhouse, however, is too weak an image, because this is a house that can kill you. It does kill people unequally, kills some people more than other people. But "scary" is a little bit like the situations of Joanna Russ's "girlchild." I love the figures of her girl-children, her growing-up stories about the older woman who rescues the younger woman, which involve a passage into maturity. Toni Cade Bambara does the same thing; she has lots of "girls coming into responsibility in a community" stories. And those are scary transitions: you become an adult, and one of the things that's involved in becoming an adult is that you know that you actually can get hurt, you actually can die, and these things are not jokes. But they're also adventures, they're part of being grown-up. So I like the idea that in some metaphorical way we are maybe becoming "adults" about technology. And that involves being a little scared, but not paranoid; and realizing that these are not devils, but they are real weapons.

CP: This is a question only Andrew can ask.

AR: I wonder whether you've been able to gauge how men read a text like the Cyborg Manifesto, especially its concluding line, "I would rather be a cyborg than a goddess." It seems to me, for example, that a certain kind of masculinist response to the manifesto, which followed all your arguments to the letter—the whole trajectory of your arguments against naturalism—might be able to conclude his reading with the following last line: "I would rather be a cyborg than a 'sensitive man.' "

DH: That's wonderful! I would rather go to bed with a cyborg than a sensitive man, I'll tell you that much. Sensitive men worry me. No, that's paranoid. [laughter]

AR: Isn't there a sense in which this is a kind of "bad girl" manifesto . . .

DH: To a certain extent, yes.

AR: because it's about pleasure *and* danger . . .

DH: and it takes a certain analogical alignment in the pornography debates . . .

AR: Right. But the "bad boy" element is a troubling one?

DH: Yes, who wants a bad boy, you don't want the masculinist response. But there's a way in which the sensitive man is the androgynous figure, the figure who is even more complete than the macho figure. And more dangerous. That's my resistance to the fact that I *do* like sensitive folks of all sexes. But the image of the sensitive man calls up, for me, the male person who, while enjoying the position of unbelievable privilege, also has the privilege of gentleness. If it's only added privilege, then it's a version of male feminism of which I am very suspicious. On the other hand, that line is written to and for women, and I think I had never imagined how a man might read it. This really is the first time I have had to imagine that line being read by people—not just male people—in a masculine subject position.

AR: There are lots of them.

DH: Yes, it never ceases to surprise me how many of them there are on the planet. [laughter] Aihwa Ong has pointed out that one very specifically American thing is to have a biological body separate from a cultural body. You find yourself in the world in a particular kind of biological body, marked with certain race, ethnicity, sexual, age characteristics, and that particular kind of marked body can, in principle, occupy any kind of subject position, but not equally easily. A male body, a male person of various kinds, could occupy the feminist cyborg subject position *and* the goddess subject position. Okay? But not equally as easily as folks who would come into a sentence like that from other histories. And the ironies would be different if you imagine yourself in such a place. Because the cyborg is a figure for whom gender is incredibly problematic; its sexualities are indeterminate in more ways than for gods and goddesses—whose sexualities are plenty indeterminate. Anyway, I like that, but I don't quite know what to do with it. I'd like to know what you do with it.

CP: Maybe it would have been better to say, "I would rather be a cyborg than a male feminist."

AR: Mmmm. Yes, well that's a different can of worms.

CP: So, *your* cyborg is definitely female?

DH: Yeah, it is a polychromatic girl . . . the cyborg is a bad girl, she is really not a boy. Maybe she is not so much bad as she is a shape-changer, whose dislocations are never free. She is a girl who's trying not to become Woman, but remain responsible to women of many colors and positions, and who hasn't really figured out a politics that makes the necessary articulations with the boys who are your allies. It's undone work.

The Actors Are Cyborg, Nature Is Coyote, and the Geography Is Elsewhere: Postscript to "Cyborgs at Large"
Donna Haraway

The Cyborg Manifesto was written to find political direction in the 1980s in the face of the odd techno-organic, humanoid hybrids "we" seemed to have become worldwide. If feminists and allied cultural radicals are to have any chance to set the terms for the politics of technoscience, I believe we must transform the despised metaphors of both organic and technological vision to foreground specific positioning, multiple mediation, partial perspective, and therefore a possible allegory for antiracist feminist scientific and political knowledge.

Nature emerges from this exercise as "coyote." This potent trickster can show us that historically specific human relations with "nature" must somehow—linguistically, scientifically, ethically, politically, technologically, and epistemologically—be imagined as genuinely social and actively relational. And yet, the partners in this lively social relation remain inhomogeneous. Curiously, as for people before us in Western discourses, efforts to come to linguistic terms with the nonrepresentability, historical contingency, artifactuality, and yet spontaneity, necessity, fragility, and stunning profusions of "nature" can help us refigure the kinds of persons we might be. We need a concept of agency that opens up possibilities for figuring relationality within social worlds where actors fit oddly, at best, into previous *taxa* of the human, the natural, or the constructed.

Inhabiting my writing are peculiar boundary creatures—simians, cyborgs, and women—all of which have had a destabilizing place in Western evolutionary, technological, and biological narratives. These boundary creatures are, literally, *monsters*, a word that shares more

21

than its root with the verb *to demonstrate*. Monsters signify. We need to interrogate the multifaceted biopolitical, biotechnological, and feminist theoretical stories of the situated knowledges by and about these promising and noninnocent monsters. The power-differentiated and highly contested modes of being of monsters may be signs of possible worlds—and they are surely signs of worlds for which "we" are responsible.

These worlds are *constructed deconstructively* by a particular "orthopedic" [1] practice that involves, among other things, an active acknowledgment of unequal personal and historical pain and a refusal of nostalgia. The split and contradictory self is the one who can interrogate positionings and be accountable, the one who can construct and join rational conversations and fantastic imaginings that change history. Splitting, not being, is the privileged image for feminist epistemologies of scientific knowledge. "Splitting" in this context should be about heterogeneous multiplicities that are simultaneously salient and incapable of being squashed into isomorphic slots or cumulative lists. This geometry pertains within and among actors.

The knowing self is partial in all its guises, never finished, whole, simply there and original; it is always constructed and stitched together imperfectly, and *therefore* able to join with another, to see together without claiming to be another. Here is the promise of objectivity: a scientific knower seeks the subject position not of identity, but of objectivity, that is, partial connection. There is no way to "be" simultaneously in all, or wholly in any, of the privileged (i.e., subjugated) positions structured by gender, race, nation, and class. And that is a short list of critical positions. The search for such a "full" and total position is the search for the fetishized perfect subject of oppositional history, sometimes appearing in feminist theory as the essentialized Third World Woman. Subjugation is not grounds for an ontology; it might be a visual clue. Identity, including self-identity, does not produce science; critical positioning does—that is, objectivity. Knowledge from the point of view of the unmarked is truly fantastic, distorted, and irrational. The god trick is to be self-identical. If the cyborg is anything at all, it is self-difference.

Feminist embodiment, then, is not about fixed location in a reified body, female or otherwise, but about nodes in fields, inflections in orientations, and responsibility for difference in material-semiotic fields of meaning. Embodiment is significant prosthesis; objectivity cannot be about fixed vision when what counts as an object is precisely what world history turns out to be about. Feminism loves another science and technology: the sciences and politics of interpretation, translation, stuttering, and the partly understood. Translation is always interpretive, critical, and partial. Here is ground for conversation, rationality, and objectivity—which is power-sensitive, not pluralist, "conversation."

Above all, rational knowledge does not pretend to disengagement, to be from everywhere and so nowhere. So, science—even technoscience—must be made into the paradigmatic model not of closure, but of that which is contestable and contested. That involves knowing how the world's agents work, how they/we are built. Science becomes the myth not of what escapes human agency and responsibility in a realm above the fray, but rather of accountability and responsibility for translations and solidarities linking the cacophonous visions and visionary voices that characterize the knowledges of the marked bodies of history. Actors come in many and wonderful forms.

Designating the networks of multicultural, ethnic, racial, national, and sexual actors emerging since World War II, Vietnamese-American filmmaker and feminist theorist Trinh T. Minh-ha suggests the phrase "inappropriate/d others." [2] The term refers to the historical positioning of those who cannot adopt the mask of either "self" or "other" offered by previously dominant Western narratives of identity and politics. To be "inappropriate/d" does not mean "not to be in relation with"–that is, to be outside appropriation by being in a special reservation, with the status of the authentic, the untouched, in the allochronic and allotopic condition of innocence. Rather, to be an "inappropriate/d other" means to be in critical, deconstructive relationality—as the means of making potent connection that exceeds domination. To be inappropriate/d is not to fit in the *taxon,* to be dislocated from the available maps specifying kinds of actors, not to be originally fixed by difference. Trinh is looking for a way to figure "difference" as a "critical difference within," and not as special taxonomic marks grounding difference as apartheid.

The term *inappropriate/d others* can provoke rethinking social relationality within artifactual nature—which is, arguably, all of nature in the 1990s. Trinh Minh-ha's metaphors suggest another geometry for considering the relations of differences among people and among humans, animals, and machines than hierarchical domination, incorporation of parts into wholes, paternalistic and colonialist protection, antagonistic opposition, or instrumental production from resource. But her metaphors also suggest the hard intellectual, cultural, and political work these new geometries will require, if not from simians, at least from cyborgs and women worried about technoscience and technoculture.

I think of my present work as an effort to ask a central question about "the promises of monsters": How may the extraordinarily literal metaphorics of the "new sciences" be productively engaged by the "new politics" of "inappropriate/d others"? Science fiction is generically concerned with the interpenetration of boundaries between problematic selves and unexpected others and with the exploration of possible worlds in a context structured by transnational technoscience. The emerging social subjects called "inappropriate/d others" inhabit such worlds. SF is also a travel literature deeply implicated in the history of colonialism and imperialism, just as it is implicated in the cultural production of the literal metaphors and poetic bodies of "high-technology" social orders. In imagination and in reality, how does SF map emerging local and global subjectivities that are constructed in the intersections of the apparatuses of production of technoscience with postcolonial, antiracist, and feminist cultural politics? I am trying to explore bodies, technologies, and fictions as they are constructed through the mediation of late twentieth-century communications sciences, in the simultaneously organic and artifactual domains of biology and medicine, industry, and the military.

SF is inherently "impure"–a major source of its lure to inappropriate/d others, for whom the "economy of the same" and its injunction to purity, textual or otherwise, rouses well-founded historical suspicions. But the "impurities" of SF are hardly utopian; they are deeply troubling as well as promising. SF is an imperialist genre, in which the "star wars" heroes riding into battle on armored dinosaurs cohabit the universe with the fantastic figures of First World feminist and multicultural imaginations. Much work needs to be done in the cultural space hinted by the intersections of science fiction, speculative futures, feminist and antiracist theory, and fictions of science. For me, the best place to locate this work remains "in the belly of the monster," that is, in the fictional and technical constructions of late twentieth-century cyborgs, site of the potent fusions of the technical, textual, organic, mythic, and political.

My present writing tries to stage conversations on the fate of the riven categories of "the human," "the natural," and "the artifactual" among heterogeneous and polyglot scientific and feminist texts. The focus is on how these discourses make possible figures of critical subjectivity, consciousness, and humanity—not in the sacred image of the same, but in the self-critical practice of "difference," of the I and we that is/are never identical to itself, and so has hope of connection to others. The tale fits together at least as well as the plot of Enlightenment humanism and science ever did, but I hope it will fit differently, negatively, if you will. My stakes are high; I think "we"— that crucial riven construction of politics—need something called humanity and nature. This is the kind of thing that Gayatri Spivak calls "that which we cannot not want." [3] We also know, from perspectives in the ripped-open belly of the monster called history, that we cannot name and possess this thing we cannot not desire. That is the spiritual and political meaning of poststructuralism/postmodernism for me. "We," in these discursive worlds, have no routes to connection other than through the radical dis-membering and dis-placing of our names and our bodies. We have no choice but to move through a harrowed and harrowing artifactualism to elsewhere.

Emerging from this process are excessive and dislocated figures that can never ground what used to be called "a fully human community." That community turned out to belong only to the masters. However, promising monsters, who are always already within, can call us to account for our imagined humanity, the pieces of which must be articulated through translation. Listening to the polyglot languages that actually, historically fill up the New World—the Americas, most certainly including the United States—I have tried to look again at some feminist discards from the Western deck of cards. The search is for the trickster figures that might turn a stacked deck into a potent set of wild cards for refiguring possible worlds. Can cyborgs and technological vision hint at ways that the things many feminists have feared most can and must be refigured and put back to work for life and not death? Located in the belly of the monster, a region of the First World in the 1990s, how can we continue to contest for the material shapes and meanings of objects, nature, bodies, and experience? How might an appreciation of a constructed, artifactual, historically contingent nature lead from an impossible but all-too-present reality to a possible but all-too-absent elsewhere? As monsters, can we demonstrate another order of signification? Cyborgs for earthly survival!

NOTES

1. Thanks to Faith Beckett of History of Consciousness for the term.

2. Trinh T. Minh-ha, ed., She, the Inappropriate/d Other (special issue), *Discourse,* 8 (Fall-Winter, 1986–87). See also her *Woman, Native, Other: Writing Postcoloniality and Feminism* (Bloomington: Indiana University Press, 1989).

3. Gayatri Spivak, lecture and discussion presented at the Center for Cultural Studies, University of California, Santa Cruz (November 1989).

Containing Women: Reproductive Discourse in the 1980s
Valerie Hartouni

San Francisco Chronicle

The Largest Daily Circulation in Northern California

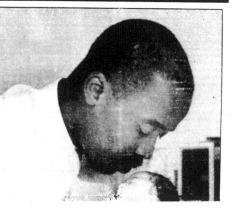

Brain-Dead Mother Has Her Baby

By Thomas G. Keane

Michele Odette Poole, weighing four pounds five ounces, belted out a healthy cry of life at 8:54 a.m. yesterday, 53 days after doctors pronounced her mother dead with a brain tumor.

Her elated father tearfully held and kissed his first daughter minutes after delivery, announcing later that the wide-eyed infant was "a carbon copy" of her mother, Marie Odette Henderson.

"Michele looks a lot like Odette," beamed Poole at a noon press conference with doctors and his attorney. "She's got lots of hair on her head. She's very alert, very active — everything seems fine."

The Richmond man shared hugs and kisses with family and friends and promised to give cigars to doc

I

"Brain-Dead Mother Has Her Baby" — so read the headline of a major West Coast newspaper in July 1986, when doctors removed an apparently healthy, thirty-two-week-old fetus from the body of Marie Odette Henderson.[1] Henderson had died fifty-three days earlier from a brain tumor; by court order, her body was kept functioning until the respiratory system of the fetus she carried had matured sufficiently to enable "independent" life. Once matured, the fetus was re-

27

Orphan Embryos Saved

Sydney, Australia

Legislators approved an unprecedented measure last night blocking the destruction of two frozen embryos and clearing the way for their adoption and implantation in surrogate mothers.

A committee of scholars had debated the fu-

ture of the embryos at the request of the Victoria state government and recommended that they be destroyed. But after a public outcry, the Upper House of the Victoria state Parliament yesterday passed an amendment allowing the embryos — produced and frozen in 1981 for an American cou-

Back Page Col. 5

moved by cesarean section and delivered into the arms of Henderson's fiancé. Shortly thereafter, doctors disconnected the woman from all life support, whereupon she was pronounced dead, again.

Henderson is not the first woman to have been maintained on a mechanical support system until the fetus she carried could survive delivery. In 1983, the "life" of another comatose, legally dead, pregnant woman was similarly prolonged until the twenty-two-week-old fetus she carried had matured.[2] And even as Henderson's doctors in California were disconnecting her respirator, their Georgian counterparts were pumping hormones, oxygen, sugar, protein, and fat through the body of yet another woman in the hopes of rescuing her fetus.[3] As of 1986, there had been at least twelve infants produced in this fashion and, given the generally favorable reception that has greeted their arrival, there is no reason to expect that others will not follow.[4]

Bizarre? Chilling? The stuff of *National Enquirer*, B-grade thrillers, or punk rock lyrics? Consider another headline, appearing in the same West Coast newspaper in October 1984, some twenty months prior to the Henderson event. Reminiscent of those following the release of the American hostages from Iran, this one read, "Orphan Embryos Saved."[5] Orphan embryos saved—embryos we do not usually think of as "orphanable," as independent life forms floating about in the world, as parentless minors, in trouble, on the loose, lost, lonely, abandoned, in need of being saved, "rescued," or adopted. To the degree that we think of them as being "in the world" at all, it is as attached and "embodied"—in a body, part of a body, and a body that is, still, necessarily and exclusively female.

That embryos might be in the world, "parentless" or "detached," "unembodied," and "unembodiable" without dramatic legal—not to mention medical—intervention suggests a dicey situation indeed, and it was into just such a situation that the embryos referred to in

H

24

the headline had apparently been cast. In vitro grown, "on ice," and awaiting implantation at the Queen Victoria Medical Centre in Melbourne, Australia, these two embryos faced what many considered an unnecessary, not to mention untimely, thaw when "parents" Mario and Elsa Rios died suddenly in a flying accident. Following their deaths, an intense legal skirmish erupted over disposal of the embryos. Whose property were they, what was their status, the nature of their relationship to each other and their "genetic sponsors," the extent of their claims? Should they be thawed and flushed, used for experimentation, or "put up for adoption"? If "adoptable," were they to be separated or kept together? And once adopted, implanted, and born, did they have rights in the distribution of their "parents' " apparently quite considerable estate?

Various courses of action were being pursued by various interested parties—Michael Rios, Mario Rios's son by a previous marriage, right-to-life groups in the United States and Australia, as well as potential surrogates from around the globe—when several more announcements were issued regarding the identity and status of the embryos. First, the infertility center at which the embryos were being stored revealed that, technically speaking, the Rios embryos were not orphans at all. They had been conceived with sperm from an anonymous donor and were thus unrelated to Mario Rios genetically; a "biological parent" was indeed alive somewhere, although one clearly uninterested in the many possible ends to which his sperm might have been put. Following this announcement, Carl Wood, head of the "infertility team" at the Melbourne center, disclosed that it was quite likely that the embryos themselves were "duds"—damaged and simply not viable "since [they] had been frozen at a very early period in the development of the freezing technique." [6] None of the embryos frozen at that time, according to Wood, had successfully implanted or developed.[7]

These details had little influence on how the issues continued to be construed and constructed publicly, as the headline "Orphan Embryos Saved" attests. Against the recommendation of a "committee of scholars," the Rios embryos were "saved" by the Victoria State Parliament when it passed IVF-related legislation regulating the procedure and providing, specifically, for the anonymous donation of "orphaned" embryos to women unable to produce eggs themselves.

Presumably, the two embryos were donated or "adopted"; about what ultimately became of them, we can only speculate.

Setting aside the particulars of each case for the moment, let us consider again both headlines: "Brain-Dead Mother Has Her Baby," "Orphan Embryos Saved." What makes these headlines make sense? Why might they seem "sensible" today when only twenty years ago they would certainly have been preposterous? What is the context, the subtext of utterances that conjoin the banal and the extraordinary in a temporarily dissonant or disruptive way? What beliefs, assumptions, and expectations allow them to be coherently rendered, taken seriously, understood as "fact" rather than "fiction"? What is the world they simultaneously construct and contain? What are the stories they tell about reproductive possibilities, relations, and relationships in late twentieth-century America, and what is the terrain they occupy and contest in that telling?

Conceptually, both headlines produce a kind of mental astigmatism; meanings temporarily blur, lose definition, appear distorted, and are resolvable only with some sort of conceptual retraining or adjustment. They require us to do conceptually, it seems, what lenses would do optically: refocus, reimage, reintegrate. But just as lenses may enable us to see the world, they also transform the world we see. And so too with these headlines. To correct the astigmatism or resolve the various conceptual distortions they present initially requires a reassembling of images and familiar categories; however, the ordered, coherent world we then "retrieve" is one we are also, through this process of reassembling, engaged in constructing. The headlines do not simply present a world that we passively perceive and assimilate, they also and significantly engage us in the making of one.

Consider again the first headline: "Brain-Dead Mother Has Her Baby." The coherence of this statement rests, in part, on a very particular understanding of "motherhood"—an understanding in which motherhood is equated with pregnancy and thereby reduced to a physiological function, a biologically rooted, passive—indeed, in this case, literally mindless—state of being.[8] Within this understanding, motherhood is cast as "natural" or "instinctual," a synonym for female, the central aspect of women's social and biological selves, the expression and completion of "female nature." It is something that just "happens," something that is initiated at conception, something that, as a biologically rooted capacity, does not depend upon a woman's consciousness for its development, but develops *as* woman's consciousness if not disturbed or thwarted in the process.

Now there are certainly other possible understandings of what "motherhood" is and entails. In the United States, conflicting claims and assumptions about its place and meaning have been at the center of many of the most fiercely waged political battles and policy disputes of the last two decades: the struggles, for example, over contraception use, abortion, surrogacy, welfare reform, divorce, custody, and day care. It is, however, only within this narrow, ideologically biologistic understanding of the term *mother* that the headline itself actually makes sense. Consider an alternative understanding, one that, for example, regards pregnancy as a biosocial experience and motherhood as a historically specific set of social practices, an activity that is socially and politically constructed and conditioned by relations of power, and that differs according to class, race, history, and culture. In this formulation, "being a mother" is not something women "are" by nature, instinct, or destiny, or by virtue of being female or pregnant. Rather, it is something women (among others) do: it is conscious and engaged work in the fullest sense of the word and an activity that is still but need not necessarily be gender specific.

The differences between this account and the first are quite dramatic, resting, as they fundamentally do, upon a distinction between social activities and meanings on the one hand and "biological processes" on the other. Were the latter account our working understanding of the concept "mother," the headline itself would be virtually unintelligible. It is not our working understanding, however, and would probably strike a fair number of the population as a somewhat extreme, certainly radical, politically interested formulation, so deeply entrenched and culturally pervasive is the assumption that motherhood is a (natural) condition, a state of bodily being rather than deliberate activity. It is this assumption—so common as to appear simply part of the "fabric of fact"—that the headline draws upon and reinforces: motherhood is something, it suggests, that simply happens and that can be sustained by mechanical means and a continuous infusion of chemicals even if there is no subject, no agent, to sustain it. The subject that knew herself as Marie Odette Henderson, after all, is dead; *she* is not present, nor for that matter is she represented except as absence or trace. Featured in the picture accompanying the headline is Henderson's fiancé, Derek Poole, cradling the infant over which he sought and won legal custody when Henderson's next of kin authorized doctors to disconnect her respirator. Indeed, to the extent that anyone actually *has* a baby, it is Poole rather than Henderson. In her present state, Henderson is, literally, a receptacle—indeed, a quite passive receptacle—for the maintenance of fetal life. She, as Aristotle might have only dreamed, is all biological process, raw reproductive material maintained by extensive mechanical and chemical manipulation. The only sense in which it could be said that *she* is a *mother* who *has* a baby is if her *sheness* is reduced to motherhood is reduced to all biological tissue and process.

Now, there are any number of other ways in which the headline could have been written, and some of these alternative constructions would have proffered a somewhat less sentimental presentation of the Henderson case. The headline might have read, "Thirty-two-Week-Old Fetus Extracted from Corpse"; "Eight-Month-Old Fetus Extracted from Dead Woman's Body"; "Deceased Delivered"; "Deceased Maintained, Fetus Successfully Retrieved"; or even "Trapped Fetus Lives." Each alternative rendering offers a more explicit account of the situation, although an account that is also more morbid, disturbing, and potentially destabilizing. What these headlines do that the original does not is foreground the third-party intervention, the hand that reaches into a surgically opened uterus and removes the fetus, the technology that permits the crossing of a hitherto uncrossable border. They make visible precisely those procedures and decisions that the original headline subsumes within the category of biological processes. They denaturalize what the original headline invites the reader to treat as normal—what is more "natural" than mothers having babies? Conversely, what is more "unnatural" than women who try and can't or women who simply won't?

What makes this particular mother having this particular baby newsworthy is that the former is brain-dead; but being brain-dead and having a baby are not necessarily mutually exclusive states when motherhood is considered a physiological "event" that expresses and completes "female nature." "Giving Life After Death"—that is how *Newsweek* ambiguously cast its headline to the story it ran on the Henderson case. And, although it is not clear who is actually "giving life"—the doctors, the courts, or a vast array of biomedical procedures—a perfectly plausible reading of the headline, and one that calms us, suggests it is Henderson, the "mother," particularly if "motherhood" is again understood in the culturally pervasive terms of the original headline. For what motherhood entails, within the sense invoked by the original headline, is commitment, dedication, and self-sacrifice, and what could be a more complete, absolute, or total expression of maternal commitment, self-sacrifice, and dedication than "giving life after death"? Henderson herself, of course, is not a conscious agent in this giving, but then neither are pregnant women more generally. Within the terms assumed by the original headline, they are merely the mediums or physical vessels for new life, not active participants in its creation or maintenance.

II

Throughout this last decade, and particularly during its first half, public discourse and debate have seemed obsessively preoccupied with women and fetuses. The decade began, need we remind ourselves, in a flurry of antiabortion, antigay, anti–ERA, profamily, prolife, pro–American rhetoric. According to the story this rhetoric told, the country had grown economically weak, militarily soft, morally decadent, spiritually impoverished, and sexually debauched: newly elected conservatives vowed a "return to basics." This entailed a full-scale assault on "affirmative action, civil rights, welfare [provisions,] and [much of the other] progressive legislation and judicial action of the previous twenty years." [9] It also entailed a mass-based crusade against liberal abortion and all that the practice marked: teenage sex, nonmarital sex, nonreproductive sex, hedonism, careerism, women's work force participation, the denigration of "traditional" gender identities, and the dissolution of the nuclear family.[10] Indeed, in the early years of the decade, abortion became not only the symbol of the general malaise that was slowly but persistently destroying the social body (as it destroyed the natural-familial-maternal body), but the ideological centerpiece of the New Right's campaign to revitalize the country politically and rehabilitate it morally.

Thus, in April 1981, a Senate Judiciary Subcommittee, headed by conservative Senators John East and Orrin Hatch, began hearings to determine the life status of the fetus. The purpose of these hearings was to bring before Congress the Human Life Statute, or S. 158—a bill that located the beginnings of "actual human life" at conception on the basis of "present-day scientific evidence" and sought to extend Fourteenth Amendment protection to that life, thereby challenging the Supreme Court's 1973 findings in *Roe v. Wade*. The Court declined in *Roe* to make a (formal) determination as to the life status of the fetus. As Justice Blackmun put it, "When those trained in the respective disciplines of medicine, philosophy, and theology are unable to arrive at any consensus, the judiciary, at this point in man's knowledge, is not in a position to speculate as to the answer." [11] The Court nevertheless established "a postnatal condition as a prerequisite for constitutional protection." [12] Maintaining that "the unborn [had] never been recognized in law as persons in the whole sense," Blackmun concluded that, within the language and meaning of the Fourteenth Amendment, the word "person" did not have any possible prenatal application. [13] Were the fetus's "personhood" subsequently confirmed, proven, or somehow "discovered," the Court seemed to imply in *Roe*, its right to life would then be guaranteed specifically by the Fourteenth Amendment, and this was precisely what proponents of S. 158 sought to establish in the context of these hearings: "that the life of each human being begins at conception" and that the Fourteenth Amendment's guarantees extend to all human beings without regard to their gestational stage.

To establish this finding, backers of S. 158 solicited testimony from embryologists, chemists, geneticists, and biologists—in short, prominent members of that community generally regarded as disinterested, beyond reproach, and able to provide, in our age, what philosophers as well as theologians provided in a more distant one: self-knowledge, knowledge of origins, knowledge of the stuff out of which we are made, but knowledge rooted in truth—objective, morally neutral, and untainted by the world. Indeed, expert after expert testified in support of S. 158, "portraying its thesis as a report from the laboratory," while colleagues and critics alike challenged their authority and expertise as scientists to make factual determinations about the beginnings of human life. [14] According to Leon Rosenberg, head of Yale's genetics department and the only dissenting voice to be included in the initial and most visible round of testimony, such determinations were beyond the purview of science. To Rosenberg's knowledge, raw, scientific data that might be used in defining human life simply did not exist. This claim was echoed by the National Academy of Science when it warned that S. 158 would not stand up to scientific scrutiny, dealing as it did with moral and religious matters or "question[s] to which science [could] provide no answer." [15]

Despite these objections and others, including the charge that S. 158 circumvented avenues provided by the Constitution for reversing Supreme Court interpretations and thus represented an attempt to exercise unconstitutional powers,[16] the Senate Judiciary subcommittee concluded, on the basis of the "evidence" before it, that "science" had indeed "demonstrated" the presence of human life at conception. The fetus emerged from these hearings a "person," but one without constitutional protection and thus vulnerably situated in liberalism's mythic state of nature, where, even for the most clever postnatal entity, life is at best inconvenient and at worst solitary, poor, nasty, brutish, and short.[17]

Although the Human Life Statute and its subsequent incarnations were not ratified, these hearings were quite successful in initiating a process of public redefinition and setting its terms. It is true that the meaning of the Fourteenth Amendment's operative word *person* was not formally broadened to include fetuses. Profoundly discredited, however, was the Supreme Court's interpretation of that meaning, a discrediting accomplished largely by members of that community popularly regarded as trafficking in "truth" and having a corner on "objectivity." [18] Similarly, while efforts to establish a strictly scientific basis for determining fetal personhood were greeted with considerable skepticism, as little more than a dubious political ploy, in the years following these hearings invocations of and appeals to "scientific knowledge," "evidence," and "advances" have become quite commonplace in the dispute over fetal personhood, and not just by those pressing for abortion's recriminalization. Indeed, with these hearings, "science" entered the political mainstream as a reservoir of nonpartisan truths about reproductive relations and has since become perhaps *the* most important ingredient contributing to the transformation of popular perceptions of the fetus.[19]

Initiating the decade, then, are legislative efforts to redefine constitutional language, broaden the meaning of the word *person*, and give concrete reality to the idea of "fetus as person." These legislative efforts have a potent analogue in popular discourse with the appearance of strange and fantastic images of fetuses in bus terminals and public restrooms, as well as on billboards, the covers of magazines, and the evening news. These images present a prenatal entity with seemingly translucent skin, suspended in empty space or floating free, vulnerable, autonomous, and alone, sucking its thumb in some representations, raising its hands beseechingly in others. A written text usually accompanies the portrait and might pose a question ("Aren't they forgetting someone?"), make an assertion ("Unborn babies are people too."), or issue a call to action ("Stop the Killing!"). Now, what this is supposed to be an image of seems obvious, and it does not appear particularly chimerical or implausible until one stops to consider that no fetus (or, for that matter, image) simply floats, alone, in empty public space, unconnected, self-generating, and self-sufficient. Moreover, no fetus (or image) is self-evidently what it is, thus raising the obvious question: What or who exactly is this? Is it the "natural man" of liberalism's mythic past extended to gestation, independent, alone, but threatened? Is it the "rugged individual" of Reagan's mythic present, abandoned by liberal administrations that could not keep their priorities straight? Is it the explorer or spaceman of the mythic future? Or is it Henry Hyde and other members of Congress who have taken to identifying themselves as "postnatal fetuses"—in Hyde's case, a 653-month-old "postnatal fetus"?

All of these allusions bring this image to life early in the decade. Pivotal, however, in giving the "still life" real life, its "own" story, was the 1984 release of *The Silent Scream*, a video production that invited the American public to witness an abortion from the "victim's" point of view. The film "purports to show a medical event, a real-time ultrasound imaging of a twelve-week-old fetus being aborted." [20] What the viewer sees are shadowy, black-and-white images, interpreted and explained by ex-abortionist physician Bernard Nathanson. Nathanson begins the narration by positioning a baby doll next to the shadow he calls a fetus: "The form on the screen, we are told, is the 'living unborn child,' another human being indistinguishable from any of us." [21] Placid initially, the "movements" of Nathanson's "unborn child" become, by his account, ever more frantic and anguished as the aspiration begins. "Sensing danger," he tells us, it retreats from the aspirator and attempts, in ever more desperate motions, to escape from its sanctuary turned tomb. It struggles violently with its arms and, seconds before it is dismembered and destroyed, throws back its head in what Nathanson characterizes as a "silent scream."

The Silent Scream was made following a study that appeared in 1983 in the *New England Journal of Medicine*, authored by bioethicist Joseph Fletcher and physician Mark Evens.[22] Fletcher and Evens noted that ultrasound imaging of the "fetal form" tended to foster among pregnant women a sense of recognition and identification of the fetus as their own, as something belonging to and dependent upon them alone. Constituting the stuff of maternal bonding, "the fundamental element in the later parent-child bond," such recognition, Fletcher and Evens claimed, was more likely to lead women "to resolve 'ambivalent' pregnancies in favor of the fetus."[23] Within the context of their study, based, significantly, on only two, entirely unrelated interviews, early ultrasound imaging appeared, to these authors, to encourage fewer abortions and thus to promote pregnancy. According to Rosalind Petchesky, it was this conclusion that apparently captured the imagination of Bernard Nathanson and the National Right-to-Life Committee and precipitated the production of *The Silent Scream*. The video, Petchesky observes, "was intended to reinforce the 'visual bonding' theory"[24]; not only does it enable us to "experience" and view the fetus as though it existed outside the body, already in relationship with the world, it also allows and encourages us to treat it as such, as a discrete but vulnerable entity in need of care.

The use of ultrasound for monitoring early fetal development as well as labor has become increasingly regularized in the United States — well over one third of all women can expect to undergo the procedure at some point in the course of their pregnancies, and, were health insurers to underwrite its routine use, many speculate that its already widespread application would expand even more. This may account for why a video such as *The Silent Scream* is "believable," why the fantasy it presents as "medical reality" is so readily accepted as "objective" and "factual" rather than contrived: paraphrasing Petchesky, the live fetal image of the clinic appears simply to have been transported into everyone's living room.[25]

However, also contributing to the credibility or apparent "facticity" of the video, and, more generally, to the transformation of popular perceptions of the fetus, have been extraordinary advances in the area of neonatology — the emergence of new methods of prenatal diagnosis and in utero treatment as well as the development of ever more sophisticated means for maintaining fetal life outside the body at earlier and earlier stages in its development.[26] Not only have these new techniques and treatments made "the fetus more accessible to the world at large, visibly, medically and emotionally," [27] they appear, in Nathanson's words, to show "beyond reasonable challenge . . . the specifically human quality of its life." [28] Physician Michael R. Harrison puts the issue this way:

> The fetus could not be taken seriously [we might ask by whom] as long as he remained a medical recluse in an opaque womb; and it was not until the last half of this century that the prying eye of the ultrasound (that is, ultrasound visualization) rendered the once opaque womb transparent, stripping the veil of mystery from the dark inner sanctum, and letting the light of scientific observation fall on the shy and secretive fetus. . . . Sonography can accurately delineate normal and abnormal fetal anatomy with astounding detail. It can produce not only static images of the intact fetus, but real-time "live" moving pictures. . . . The sonographic voyeur, spying on the unwary fetus finds him or her a surprisingly active little creature, and not at all the passive parasite we had imagined.[29]

Indeed, as Dr. Frederic Frigoletto, a pioneer on the frontier of fetal therapy describes it, observing the fetus in utero is "almost like going to a nursery school to watch the behavior of three-year-olds." [30] No longer a "medical recluse" or a "parasite," the fetus has been grasped as an object of scientific observation and medical manipulation, not to mention anthropomorphic imagination. Thus, although the various new techniques of therapy and repair have not yet made viable the construction of "fetus as person," they have made quite commonplace the construction of "fetus as patient," an entity requiring a separate physician and often separate legal advocate, or, following Harrison, "an individual with medical problems . . . [who] can not make an appointment and seldom ever complains." [31] Harrison's choice of words and Frigoletto's comparison are instructive. The fetus is neither a "person" nor an "individual" in the strict sense of either word or, as we have seen, for constitutional purposes. However, both descriptions also make clear that the distinction between "patienthood" and "personhood" is not an easy one to maintain, and the ground upon which it rests has been eroding rapidly.

III

If we have witnessed a growing public preoccupation—indeed, obsession—with fetal life over the course of the last decade, no less obsessive has been the popular preoccupation with women. A special report on abortion aired by ABC's *Nightline* framed the matter and, one could say, the decade, this way: "With new technologies peering into the womb, women have been forced to peer into their hearts." [32] Although it is hardly novel to present abortion as a conflict between the opposing claims of fetuses and women, the opening statement of *Nightline*'s special report is striking in that it reformulates this opposition and, in so doing, fundamentally transforms it. The conflict it constructs is not between fetuses and women, but rather between "truth" and "desire": on the one side is "truth"—technology, objective observing, the womb as a thing in itself and the site of self-evident but only recently deciphered meanings, the uncovering of knowledge through scientific investigation; on the other side is "desire"—women, self-reflection, the heart as the site of moral (maternal?) meanings, the recovery of knowledge through introspection. Given the traditional ontological and epistemological valorization of the first term of this truth-desire antinomy, a "conflict" in these terms is clearly no conflict at all. It is, at most, a confrontation the outcome of which is foregone: truth must always overcome desire, desire must always give way. Rendered invisible—indeed, irrelevant—by this formulation is some of the dispute's most fiercely contested ground. Rendered derivative and positioned peripherally to their bodies are women.

Just how truly chimerical this antinomy is becomes clear once its first term, *truth*, is situated and contextualized. Technologies are peering, *Nightline*'s report begins, but technologies, we know, do not themselves "peer"; they are instruments and relations that facilitate or obstruct but, above all, construct "peering." Likewise, "peering" is not itself a benign, impartial, disinterested, or disembodied activity, but is both mediated and situated within interpretive frameworks, points of view, sets of purposes. The question, then, is who is peering, what are they looking for and why, with what predispositions, assumptions, expectations, and predilections? Someone (a physician?) is looking and looking for some "thing" (the fetal patient?), for some purpose (diagnostic?), by "consent" or with the cooperation of the woman in whose body the womb is situated.[33]

And then there is the womb itself, the "object" being peered into, and the commonplace knowledge that fetuses, when they hang out, tend to hang out in wombs—this is (still) a condition of being a fetus, popular representations notwithstanding. By peering into a pregnant woman's uterus, one can expect to encounter at least one fetus, perhaps more. So, exactly what new mystery has suddenly been unveiled? What is it that the "new technologies" are now "encountering" that "they" weren't, in some sense, already prepared to encounter? What is being "found" that isn't also being looked for? When fetal pioneers Harrison and Frigoletto "peer," they greet, as we have heard them describe their sightings, "shy, secretive" but "surprisingly active little creatures," behaving like three-year-olds at nursery school. Placing the content of the once "opaque womb" "under the light of scientific observation," they have apparently discovered thirty-six-month-old prenatal toddlers. When Representative Henry Hyde, among other staunch opponents of abortion, turns his gaze to the human world, he sees "postnatal fetuses"; when "science" turns an "impartial" eye to the content of the uterus, it now sees "prenatal toddlers." The gaze of Hyde is generally recognized as "interested"; the gaze of science, disinterested. On close inspection, what really distinguishes them?

Situated in opposition to the supposedly unsituated gaze of technology is the second term of the antinomy, the truly situated gaze of women, directed not toward the uterus, but toward the heart and the heart as it exists, not in some physiological sense, of course, but metaphorically. *Heart* has its etymological origins in the Greek term *kardian* or *cura*, indicating care or concern, and while its metaphorical associations in the West have varied historically, it has enjoyed a consistent identification as the decisive center of things, the source of multiple truths, the site of moral sensibility, moral conscience, and even unwritten law.[34] Located within the heart or buried in its depths is a particular kind of knowledge of life's interiors and rhythms or things primal, vital, and sacred—a knowledge that is clearly available to all, but traditionally considered more fully and easily ascertained by women. Indeed, in things of the heart both moral and affectional women have traditionally been regarded as the more literate gender, by virtue of their generative capabilities and their intimate, "primordial" relation to the production and reproduction of life.

Especially interesting, therefore, is *Nightline*'s counterposing of the uterus and the heart, given that women's "special" ability to read the heart and coherently render its teachings has conventionally been linked to their uteruses, their ability to create life and sustain it. If women must be "forced" to peer into their hearts, forced to do what once took place "naturally," something has ruptured an organic system of communication, of "knowing" and "doing"; something is clearly awry and that "something" appears to be legalized abortion. Liberalized abortion is commonly depicted within popular discourse and debate as having rendered women, if not dyslexic, then illiterate in affairs of the heart—(unnaturally) uncaring and selfish as well as emotionally and physically hostile to the unborn.[35] In this respect, "peering" technologies function as remedial aids. By exposing the interior life of the pregnant uterus, they enable women to reintegrate "knowing" and "doing" (through visual identification and bonding) or once again to "read" all that is believed to be inscribed on their hearts and act accordingly.

Women's alleged "illiteracy" in matters of the heart has been the focus of much anxious public attention this decade. Despite their obvious benefits, technological developments such as in utero therapies and sonogram have been deployed to expose its scandalous extent and thus to perpetuate "the most deadly anti-woman bias of them all, namely, that unless women are carefully controlled, they will kill their own progeny."[36] The images, not to mention the rhetoric, have been powerful and, on some level, clearly persuasive. In April 1982, for example, President Reagan attributed the existing severe unemployment rate to the large number of women entering the work force.[37] Several months later, "profamily" activists claimed their first victory with the defeat of the Equal Rights Amendment. Underpinning Regan's otherwise forgettable reading of the troubled economy and at the center of anti–ERA ideology sat the assumption that women's essential identity could be fully realized only in the home, as housewives and mothers. By seeking paid work outside the family, "follow[ing] the siren call of women's liberation . . . , and competing in the labor force for scarce available jobs," women were said to be not only jeopardizing the family's stability (as well as the nation's), but denying their natural, God-given destiny to bear children and rear them.[38] Just how inescapably perverse or unnatural women's participation in the work force is was graphically established, according to prolife forces, by the horrifying costs such participation exacted yearly: quite contrary to their "innate" maternal drives, women were aborting their unborn. Just as some animals kill their offspring when they are disturbed or in some way confused, so too were women killing theirs, in "restless agitation against a natural order." [39]

"Hard" evidence could hardly be marshaled to support these clearly dubious claims. However, innuendo and insinuation in the context of what were largely successful efforts to breathe real life into the still life of the fetal form have proved every bit as effective in providing a convincing and compelling portrait of women as out of control and maladjusted, not to mention miserable, agitating against nature, and thus unable to "read" its directives accurately. Consider, in this regard, the development of fetal protection statutes. In 1982, the same year the president told American women to go home, a small group of obstetricians and geneticists declared that "medicine [was] far enough along for them to start treating fetuses as patients." [40] With this declaration came the rapid expansion of efforts not only to treat the "fetal patient" but to protect it from potential abuse or neglect. Juvenile courts subsequently began to assume jurisdiction over the content of pregnant women's wombs, and right-to-life efforts to rescue fetuses through forced medical intervention or incarceration of the women carrying them were intensified and highly publicized. Increasingly subject to legal scrutiny and criminal prosecution were pregnant women who smoke, drank, had sex, ingested legal as well as illicit drugs, refused major surgery (for example, cesareans), failed to follow the advice or instructions of their physicians, failed to obtain adequate prenatal care, worked or lived in proximity to teratogenic substances, or engaged in any range of activities deemed "reckless" and potentially detrimental to fetal life.[41]

Statutes currently invoked to monitor or bring under critical scrutiny the conduct of pregnant women were originally designed to protect pregnant women, to provide criminal sanctions against an assailant when assault and battery resulted in the death of the fetus.[42] Expanded to reflect new prescriptions for fetal health and vigorously lobbied for by antiabortion organizations, however, the focus of these statutes has shifted to reflect a substantially different agenda. Primarily concerned not with the harm done *to* women but with the harm done *by* them, these statutes now render the former "victim" a potential assailant (or aborter).[43] Depicting a woman and the fetus she carries as two separate, adversarily related individuals—the one a potential killer, the other innately innocent—they engender and promote the notion that, whereas women once nurtured their unborn, they now regularly abuse or neglect them and cannot be trusted not to. Where gestation was itself once the most natural of processes, it has now become treacherous. Additionally reinscribed with these statutes is an assumption as old as it has been intractable, that women are merely vessels, "containers" that can be "opened" in the name of fetal health even if such intervention places their own lives and health at stake.[44]

That such statutes both entail and lend legitimacy to the drastic abridgment of such fundamental rights as privacy and bodily integrity seems obvious: the American legal system has traditionally refused "to recognize the right of anyone—born or unborn—to appropriate the body of another for his or her own use."[45] However, in gendered legal discourse, the female body has rarely been considered a "body" in some general, unmarked sense; to the contrary, the American legal system has traditionally subsumed the female body to the maternal body.[46] Now, where gestation is considered the natural fact of female bodies and the essential core of women's identity, where all women are regarded as "mothers," actual or potential, questions of appropriation are largely irrelevant: a body realizing its destiny can hardly be regarded as a body under siege, a body "appropriated."[47] When such questions do become relevant, therefore, as in the case of fetal protection statutes, they are readily configured as a problem of deviance—of reckless conduct or instincts gone awry. Indeed, that appropriation becomes a question at all reveals a tear in the fabric of fact and a fracturing of the core of women's identity. Fetal protection statutes, then, mend and solder. They reassert the "voice of the natural" where this voice has been muted or silenced and thereby function, perhaps most importantly, as an indictment. For inasmuch as it is through women that this voice has traditionally spoken, fetal protection statutes imply by their mere existence that women have lost heart or touch with the deepest source of their identity and thus have become not only dysfunctional but potentially dangerous.

But fetal protection statutes tell only one story among many. If these statutes conjure an image of women stung with a certain Dionysian madness, that image comes to be mirrored and reinforced early in the decade by another, ostensibly more forgiving one: women undergoing abortions only to emerge from their "madness" to a knowledge of the unmistakable horror of what they have done and become.[48] Early in the decade, there "emerges" a "condition" from which it is said an ever-increasing number of women are suffering—a condition that comes to be known as postabortion syndrome. By the decade's end, so pervasive will this "condition" appear to have become that Surgeon General C. Everett Koop will recommend committing up to $100 million for its research[49]—a move that mystifies by medicalizing what seems clearly a "discourse-generated" malady, the spread of which is attributable less to the recovery of lost identity than to the alleged "discoveries" of peering technologies, the public proliferation of arresting representations of fetal life, and the equally arresting representations of women taking that life capriciously, casually, and selfishly.

Postabortion syndrome narrates the potentially severe physical, psychological, chronic, and debilitating consequences of women's seduction and debauchery by abortion providers, the media, and feminism. It tells a story of subversion, moral laxity or turpitude, and self-mutilation, of women losing a sense of their nature and place, and of their subsequent victimization. But it is also a story of redemption, for what fetal protection statutes achieve through law, postabortion syndrome rights through female suffering: "nature" ultimately reasserts itself, through conscience and "absence." Consider a few of the syndrome's vast range of indications: guilt, remorse, despair, unfulfillment, withdrawal, helplessness, decreased work capacity, diminished powers of reason, anger and rage, seizures, loss of interest in sex, intense interest in babies, thwarted maternal instincts, residual "motherliness," self-destructive behavior, suicidal impulses, hostility, and child abuse.[50] Clearly reinscribed here as "health" are fairly conventional codings of women's bodies and lives. But beyond a necessarily inferential account of both the disruption and restoration of some sort of natural order, what postabortion syndrome seems most importantly to provide are reassurance and consolation. Indeed, this may explain why it has so captured the public imagination and the imagination of certain public health officials: exposed in the absence of something to be mothered are maternal instincts or, in Koop's words, the longings and aspirations of "motherliness"; motherhood is reiterated as women's "true" desire and interest as well as innate need.

Embellishing this picture and embroidering its tragic qualities is the "emergence" of yet another "condition" presented as having already reached frightening proportions by the time it enters popular culture and consciousness midway through the decade: the "epidemic" of infertility. To the images of women killing for reasons of convenience, abusing their "prenatal toddlers" through reckless conduct and neglect, and suffering inconsolably from having "stomped out" life only to realize its "true" nature and theirs, the recent discourse on infertility adds still other images: to the socially deviant and the psychologically deranged, this discourse conjoins the biologically deficient. Each is a face of women's agitation against the natural order, but the special contribution of the infertility discourse lies in its provision of "evidence," scientific evidence, that the ultimate goal of womanhood is motherhood, that who and what women are is fundamentally and inalterably rooted in their reproductive capacity. If there is any doubt about what women should find upon peering into their hearts, this discourse quells them.

Profiled in the popular press are accounts of women who believed they could "have it all," who "followed the siren call of women's liberation" and, by their own account, "lost all [their] womanness" in the process.[51] Unable to procreate and finding themselves driven by what fertility specialists characterize and treat as an innate need to do so, these women have been reduced to desperate caricatures of their former self-assured selves, struggling with feelings of inadequacy and failure.[52] Also prevalent among these women are feelings of betrayal, guilt, and remorse, for what the popular press makes unmistakably clear is that the present "epidemic" of infertility is largely the result of "life-style." According to *Time* magazine, for example, primarily responsible for the "epidemic" are individual choices and attitudes: delayed childbearing, liberalized and pluralized sexual behavior, multiple abortions, pelvic inflammatory disease, sexually transmitted diseases, strenuous athletic activity such as dancing, jogging, and distance running, and high-stress corporate/professional employment.[53] Echoing Aristotle and Hippocrates, *Time*'s message is not particularly subtle: women's loss of biological function is linked to their denial of this function through the pursuit of nontraditional career paths and other activities. The natural expression or ultimate goals of female instincts are marriage, motherhood, and home; clearly unnatural and even "dangerous" are abortion, delayed child-bearing, nonprocreative sex, and women's work force participation. Resurrecting the adage "Anatomy is destiny," and giving it renewed force, Patrick Steptoe, "father" of the first in vitro conceived or test-tube baby, framed the matter this way: "It is a fact that there is a bio-logical drive to reproduce. Women who deny this drive, or in whom it is frustrated, show disturbances in other ways." [54]

What *Time* and other popular newsmagazines fail to report are precisely those details that expose the apparently seamless and self-evident reality of infertility as a contestable composition of specific and predictable interests. For example, contrary to its popular pre-sentation as a problem that overwhelmingly afflicts white, affluent, highly educated women, incidences of infertility are actually higher among the nonwhite and poorly educated. Black women are one and a half times more likely to be infertile than their white counterparts, but it is not black women whom specialists seek to save from their own nonfunctional bodies or to assist in fulfilling their biological destinies.[55] When *Life* magazine chooses to feature the issue of infer-tility along with the various new technologies that have been devel-oped to "treat" it, on its cover is situated not a black baby or a brown one, but a pink, blue-eyed toddler.[56] What world is being held in place, however precariously, by the development and use of these new technologies is hardly a mystery. Both text and subtext are straightforward: white women want babies but cannot have them, and black and other "minority" women, coded as "breeders" within American society (and welfare dependents within Reagan's America), are having babies "they" cannot take care of and "we" do not want.[57]

Minus its deeply conservative pronatalist and racist agenda, infer-tility thus features a demographic profile considerably different from that suggested by popular discourse. But also brought into sharp re-lief when this agenda is set aside is a collection of "causes" radically more disturbing than "life-style" trends. Although infertility is gener-ally rendered a problem that afflicts only women in significant num-bers, it is the case that men are as likely as women to be infertile, and that incidences of infertility for both are attributable to a diverse

combination of social and iatrogenic factors: environmental pollution, exposure to hazardous toxins in the workplace, and unsafe
working conditions in addition to inadequate health care, misdiag-
nosis of disease, and sterilization abuse. While delayed childbearing
contributes to low fertility rates, it is hardly the only or even the primary ingredient generating this "epidemic."

And, finally, as for the "epidemic" itself, the actual rate of infertility
in the United States has remained relatively constant over the past
two decades.[58] What has changed is the expansion of possibilities for
treatment, at least so far as women are concerned. Within the last decade, new technological options have been developed that allow
women to "overcome" what fertility specialist Carl Wood describes
as "nature's defects." [59] While still experimental and of disputed effectiveness, these techniques have nevertheless been aggressively
promoted since their introduction, and much about the sudden appearance of an "epidemic" seems clearly related to their marketing
not only to the involuntarily childless but to funding and regulatory
communities as well.

What each of these accounts offers is the beginning of an alternative reading of infertility that challenges its construction and representation within popular culture not only as a "disease" primarily af-
flicting affluent, highly educated, married white females, but as
evidence of a disruption in the natural order of things largely attributable to female deviance. Seen from other angles, infertility cuts a
considerably different form. To the extent that it is a problem, that
problem is configured by racism, corporate greed, and the effects of
socially stratified health care provisions. To the extent that it contains
a story, that story joins others in rehearsing the social and physiological ravages of late capitalism. However, through its incorporation
into the political-medical discourse on the dangers of denatured
women and the disfigured condition of motherhood in a postlibera-
tion era, most of the important details of infertility's causes and victims are totally eclipsed. Within the constraints of this discourse and
juxtaposed with another of the decade's dominant images, free-floating or dismembered fetuses, infertility functions as a condensed symbol of the consequences of "womanhood" not kept to its natural
place.

IV

In February 1989, "after consulting experts in fields ranging from de-
mography to industrial design to medicine to interstellar travel," *Life*
magazine offered its readers a tour of life in the twenty-first century
titled "Visions of Tomorrow." Dividing the journey into three
phases—self, society, and planet—*Life*'s cover story opens with a
fourteen-inch, two-page glossy photo of a disembodied uterus, cap-
tioned, "MOTHERS OF INVENTION: New Ways to Begin Life." Situated on a steel table, under operating room lights, with tubes extending out from both sides carrying saline solution to the implanted
embryo, the out-of-body uterus is part of a series of experiments on
fertility conducted by a team of Italian physicians in 1986. Although
subsequently halted on ethical grounds, their research, *Life* tells us,
"is expected to be resumed," and to culminate, as the photograph
itself suggests, in the production of an artificial womb or placenta.
The text continues, "By the late 21st century, childbirth may not involve carrying [embryos or fetuses to term] at all—just an occasional
visit to an incubator not so different from the one shown at left.
There, the fetus will be gestating in an artificial uterus under conditions simulated to re-create the mother's breathing patterns, her
laughter and even her moments of emotional stress." [60]

The incubator to which this text refers is not, strictly speaking, an
artificial apparatus replicating the environment of the womb; it is an
impregnated uterus, the womb itself, excised from a body and me-
chanically maintained. We could call it an "incubator," a model of
things to come, as *Life* does, but why then would we call Marie
Odette Henderson (the "brain-dead mother [who had] her baby")
"mother"? What distinguishes the uterus on the steel table from the
legally dead pregnant woman who was maintained by a life-support
system until the fetus she carried had sufficiently matured, her "maternal contribution" simulated by means of a gently rocking bed, music, and the voices of nurses?[61] Indeed, why is the former declared by
some to be ethically monstrous, dangerous, or threatening, and the
latter presented as an exquisite moment in the history of maternal
self-sacrifice? What exactly is the difference between a disembodied
uterus gestating and an embodied uterus when the body that contains it is dead? Both are "incubators," but about each a dramatically
different story is told. *Life*'s rendering of reproduction is situated in

some distant future and is about bold new beginnings, mastery over the treacherous inconstancy of nature, and fundamentally reconstituting the human world. According to the text accompanying the photograph, "It will demand as much new thinking in religious and moral terms as it will in medical." [62] But *Life*'s rendering is also about the needs of the infertile, the unborn, and the prematurely born—it is partly in response to these sets of interests that this new technology is being developed. By contrast, the story told through Henderson is of things lost or slipping away, not simply a life, but a world in which mothers were mothers no matter what. Marie Henderson, dead, is counterpoint to the decade's discourse on women, the mirror opposite of her living counterpart who callously takes or abuses life. In the story told through Henderson, the woman dies, but fetus and mother survive. Nature prevails—mother nature, the nature of mothers—even as nature (death) is overcome.

Now, the interesting thing about these two stories is that each contains and constructs the other. The world held in place by Henderson as brain-dead mother, the world of conventional gender meanings and identities, is precisely the world the new technologies of reproduction destabilize; and, paradoxically, the world these new technologies destabilize is also reinscribed by their ostensible purpose and use as well as by the stories they appear to tell. What, after all, currently legitimates the development of reproductive technologies? They are said to help women realize their maternal nature, their innate need to mother. And what happens to women if this need is denied or frustrated? According to fertility specialist Patrick Steptoe, "They show disturbances." Yet, even as "maternal nature" is asserted, the certainty of what it is and the ability to render it coherently are profoundly shaken. Is Henderson "mother," incubator, or "prototype"? Or consider "surrogate motherhood," where the desire to conserve "nature" and "tradition" results in their utter disruption. As ethicist Arthur Caplan notes about the "Baby M" trial: "This case chips away at our certainty of understanding our concept of mother. It's socially and culturally disturbing. It takes away a reference point." [63] Is it really any wonder that courts and legislatures across the land moved so quickly to outlaw the procedure while fertility specialists began to see in the development of the artificial uterus a viable, ethically preferable alternative?[64]

There are, of course, still other conceptual incoherencies and cultural disturbances generated by the application of new reproductive technologies and techniques. There are free-floating fetuses said to be in need of protection from their "natural carriers," "more helpless . . . than the tenant farmer or mental patient," [65] and somehow safer outside the womb than in it. There are prenatal toddlers and postnatal fetuses, embryos on ice and "orphaned," frozen embryos apparently so numerous now as to constitute a "national population," and frozen embryos at the center of an increasing number of intense legal skirmishes prompted by divorce. Are these embryos "joint property or offspring?" asked a recent headline regarding a particularly acrimonious case in Tennessee. "He" maintains they are joint property and does not want his wife "to attempt to bear his child" using the embryos after they are divorced. As he sees it, he has "an absolute right to decide whether or not he will become a parent." [66] She contends that they are living things, "preborn children," offspring. Indeed, as she sees it, she has more rights "as the mother" [67] and seeks "to rescue her babies from the concentration camp of the freezer"—where, it must be remembered, she herself interned them.[68] According to her lawyer, hers "is the yearning of a woman to be a mother" [69]—except, of course, "natural motherhood" presupposes a world of male prerogative, one in which his rights are, as he claims, absolute.

Here, as in cases sketched elsewhere in this essay, invocations of natural rights and relations, maternal yearnings, and paternal prerogatives all draw upon, conjure, and attempt to reconsolidate a pregiven, properly ordered world of fixed meanings and identities. Indeed, such invocations aim to recover and stabilize what is perceived as having become temporarily decentered, and in this they are clearly and blatantly reactionary. However, in their move to restabilize, invocations of the natural also betoken a certain political possibility: contradictions that cannot be resolved or suppressed, a fracture in the dominant discourse that suggests a dramatic weakening of its strength, a disruption in privileged narratives that renders them highly vulnerable to contestation. At such moments, a space opens for the entry of new voices and political alliances as well as new interpretations, understandings, and conceptualizations of practices that are a site of social conflict but whose meanings and organization are by no means given.

It is difficult to see in the regressive and reactionary character of this decade's discourse on reproduction anything remotely resembling alternative, liberatory political possibilities. Throughout much of the decade, those of liberal, left, and feminist political commitments have together scrambled merely to defend "women's rights" against a continuous onslaught of legislative, judicial, and medical incursions and have been forced, moreover, toward a range of narrowly construed and often confused political responses and settlements. Within the terms, boundaries, categories, and codings of this decade's discourse on reproduction, political possibility has indeed seemed drastically circumscribed. But read differently, read symptomatically, the sheer and concentrated attention directed throughout the decade toward disciplining and managing women's bodies and lives—the fierce and frantic iteration of conventional meanings and identities in the context of technologies and techniques that render them virtually unintelligible—signals, among other things, the profound instability and vulnerability of privileged narratives about who and what "we" are. What looks like the narrowing of possibilities from one angle betokens their presence and proliferation from another.

Contained in the disruption of conventional meanings and identities and their particular vulnerability to contestation are numerous possible political openings—multiple points of resistance as well as projects of reconstruction. Naming and seizing these possibilities, however, require imagination, a new political idiom, as well as a certain courage—to eschew a lingering attachment to things "natural" and "foundational," and to jettison the essentialism clung to by all extant participants and opponents of the repro-tech drama. It requires the courage to take seriously the socially and technologically produced opportunity to invent ourselves consciously and deliberately, and in this to develop the practical, political implications of the philosophical claim that "we" are only and always what we make.

NOTES

1. *San Francisco Chronicle* (July 31, 1986).
2. *Newsweek* (April 1, 1983), 63.
3. *Newsweek* (August 11, 1986), 48.
4. Ibid. Following the extraction of a viable fetus from the body of a brain-dead woman, Dr. Russell K. Laros, Jr., had this to say: "The experience left me with real confidence that this can be done without any great difficulties. . . . In the future, I'll suggest

to family members that the option is there." Cited in Gena Corea, *The Mother Machine* (New York: Harper & Row, 1985), 281.

5. *San Francisco Chronicle* (October 24, 1984).

6. Peter Singer and Deane Wells, *Making Babies: The New Science of Conception* (New York: Charles Scribner's Sons, 1985), 87.

7. Ibid.

8. On this understanding—from the obvious to the obtuse—there of course exists a rich and ever-expanding feminist literature.

9. Rosalind Petchesky, "Abortion, the Church and the State," unpublished manuscript, 15; Zillah Eisenstein, "The Sexual Politics of the New Right: The 'Crisis of Liberalism' for the 1980's," *Signs*, 7, 3 (1982), 567–88.

10. Rosalind Petchesky, "Antiabortion, Antifeminism, and the Rise of the New Right," *Feminist Studies*, 2 (1981), 206–46.

11. *Roe v. Wade*, 410: U.S. 113 (1973), 159.

12. Gary Jacobsohn, *The Supreme Court and the Decline of Constitutional Aspiration* (Totowa, N.J.: Rowman & Littlefield, 1986), 131.

13. *Roe v. Wade*, 157–59, 161–62.

14. Daniel Wikler, "Abortion, Privacy and Personhood: From Roe v. Wade to the Human Life Statute," *Abortion: Moral and Legal Perspectives*, ed. Jay L. Garfield and Patricia Hennesey (Amherst: University of Massachusetts Press, 1984), 252.

15. Resolution of the National Academy of Science, April 28, 1981, cited in Laurence Tribe, "Prepared Statement before the United States Senate on the Human Life Bill," *Abortion, Medicine and the Law*, ed. J. Douglas Butler and David F. Walbert (New York: Facts on File, 1986), Appendix 2, n. 5.

16. Ibid., 481.

17. "For where-ever any two Men are, who have no standing Rule, and common Judge to Appeal to on Earth for the determination of Controversies of Right betwixt them, they are in the State of Nature, and under all the inconveniences of it." John Locke, *Two Treatises of Government*, ed. Peter Laslett (New York: New American Library, 1963), II, 91. "It may be perceived what manner of life there would be, where there were no common power to fear. . . . In such a condition, there is no place for industry . . . no culture of the earth; no navigation . . . no commodious building . . . no account of time; no arts; no letters; no society; and which is worst of all, continual fear and danger of violent death; and the life of man, solitary, poor, nasty, brutish, and short." Thomas Hobbes, *Leviathan*, ed. Michael Oakeshott (New York: Collier, 1974), 101, 100.

18. A "letter to the editor" that recently appeared in a progressive paper by a woman representing herself as "pro-choice for many years" but now "pro-life" gave an account of these hearings that, while riddled with inaccuracies, is a clear testimony to their success in giving the impression that the issue of fetal personhood was indeed "solved" and in objectively defensible terms. She writes: "In congressional hearings not long after Roe v. Wade, the question of when the exact moment human life begins was explored by top physicians and biologists from around the U.S. Everyone who testified, whether pro-life or pro-choice, said the same thing—human life begins at the moment of conception. Given this statement, it is inconceivable that any group that promulgates itself as one that upholds the rights of all against exploiters could possibly support elective abortions." *In These Times*, 13, 16 (1989), 15.

19. Janet Gallagher, "Fetal Personhood and Women's Policy," *Women, Biology and Public Policy*, ed. Virginia Sapiro (Beverly Hills: Sage, 1985).

20. Rosalind Petchesky, "Fetal Images: The Power of Visual Culture in the Politics of Reproduction," *Reproductive Technologies: Gender, Motherhood. and Medicine*, ed. Michelle Stanworth (Minneapolis: University of Minnesota Press, 1987), 29.

21. Ibid., 62.

22. Joseph C. Fletcher and Mark I. Evens, "Maternal Bonding in Early Fetal Ultrasound Examinations," *New England Journal of Medicine*, 308 (February 17, 1983), 392–93.

23. Ibid., 392; Caroline Whitbeck tells of a "Boston obstetrician who subjects her patients to ultrasound imaging every month of pregnancy for the purpose of 'helping the woman bond to her baby.' " "Fetal Imaging and Fetal Monitoring: Finding the Ethical Issues," *Embryos, Ethics, and Women's Rights*, ed. Elaine Hoffman Baruch, Amadeo F. D'Adamo, Jr., and Joni Seager (New York: Harrington Park, 1988), 56.

24. Petchesky, "Fetal Images," 59.

25. Ibid.

26. These technological advances fuel the quite distorted popular impression that fetuses now capable of survival can be (and are being) aborted, legally. Able-baby survival of twenty-four-week-old fetuses is still quite rare—indeed, considerably less than 50%. Also important to keep in mind is that less than .01% of all abortions actually occur between the twenty-fourth and twenty-sixth weeks of pregnancy.

27. Gallagher, "Fetal Personhood," 92.

28. Bernard Nathanson, *The Abortion Papers: Inside the Abortion Mentality* (New York: Frederick Fell, 1983), 16–17. Indeed, so convinced was Nathanson that the various new techniques of fetal therapy and repair "revealed" the personhood of the fetus, he quit his New York City-based abortion practice and joined the crusade to recriminalize the procedure. "What persuaded me to change my mind [regarding the life status of the fetus] was the marvelous new technology which has served to define beyond reasonable challenge the nature of interuterine life, the inarguably and specifically human quality of that life. Where is the scientific evidence, where are the new developments which would serve to convince us of the contrary view, that what is in the uterus from the beginning of pregnancy is less human today than that which we perceived in 1965? We had no ultrasound in those years, no fetal heart monitoring, no fetoscopy, no invitro fertilization. Science marches inexorably towards a deeper understanding of the uniquely human qualities of the fetus."

29. Cited in Ruth Hubbard, "Personal Courage Is Not Enough: Some Hazards of Childbearing in the 1980's," *Test-Tube Women*, ed. Rita Arditti, Renate Duelli Klein, and Shelley Minden (Boston: Pandora, 1984), 348–49.

30. Ibid., 349.

31. Ibid.

32. April 7, 1989.

33. The point is not that technologies facilitate only reactionary conclusions and uses, but that the context/relations, the "who," determines in toto what is seen; clearly, with these same technologies we could see and conclude other things.

34. Jacques Le Goff, "Head or Heart? The Political Use of Body Metaphors in the Middle Ages," *Fragments for a History of the Human Body: Part Three*, ed. Michael Feher with Romina Naddaff and Nadia Tazi (Cambridge, Mass.: Zone, 1988), 13–26.

35. On the effects of liberalized abortion, see, for example, Don Baker, *Beyond Choice: The Abortion Story No One Is Telling* (Portland, Ore.: Multnomah, 1985); Ann

Saltenberger, *Every Woman Has a Right to Know the Dangers of Legal Abortion* (Glassboro, N.J.: Air-Plus Enterprises, 1983), especially Chapters 6–8.

36. Marvin Kohl, cited in Barbara Hayler, "Review Essay: Abortion," *Signs*, 4 (Winter 1979), 322–23.

37. In Reagan's own words as reported in *Time* (July 12, 1982, 23): "Part of the reason for the high unemployment 'is not so much recession as it is the great increase in people going into the job market, and ladies, I'm not picking on anyone, but [it is] because of the increase in women who are working today and two-worker families.' "

38. Phyllis Schafly, *The Power of the Positive Woman* (New York: Jove, 1977), 78, 212.

39. Interview with prolife activist in Kristen Luker, *Abortion and the Politics of Motherhood* (Berkeley: University of California Press, 1984), 159–60.

40. Gina Kolata, "Operating on the Unborn," *New York Times Magazine* (May 14, 1989), 35.

41. Rex B. Wintgerter, "Fetal Protection Becomes Assault on Motherhood," *In These Times* (June 10–23, 1987); Nan Hunter, "Fetricide: Cases and Legislation," Memorandum to the ACLU, Reproductive Freedom Project (May 5, 1986).

42. Nan Hunter, "State Legislation Concerning 'Fetricide' and 'Wrongful Birth/ Wrongful Life,' " Memorandum to ACLU Affiliates, Reproductive Freedom Project (April 15, 1983).

43. Fetal protection statues clearly affect all women of childbearing age, but operate with additional cruelty against women who lack class and/or race privilege. The very conditions of these women's lives indict them: poverty, lack of access to adequate/ quality medical care, and often toxic environmental living and working conditions constitute what the courts would call "negligent fetal injuries."

44. George J. Annas, "Pregnant Women as Fetal Containers," *Hastings Center Report* (December 1986), 14; Gallagher, "Fetal Personhood," 104–5.

45. Gallagher, "Fetal Personhood," 106. See also Dawn E. Johnsen, "The Creation of Fetal Rights: Conflicts with Women's Constitutional Rights to Liberty, Privacy, and Equal Protection," *Yale Law Journal*, 95 (January 1986), 599–625.

46. In *The Female Body and the Law* (Berkeley: University of California Press, 1988), Zillah Eisenstein appears to make a similar point and to develop similar themes.

47. While pregnant women can be forced to abstain from "unhealthy" behavior, undergo blood transfusions, or have major surgery for the good of the fetus, courts would never coerce an adult male (or female) to undergo, for example, a bone marrow transplant for the good of son or daughter, mother, cousin, brother, or friend, even if he (or she) happened to be the only compatible donor. In adjudicating such a situation a Pittsburgh judge had this to say: "Morally, this decision rests with the defendant and, in the view of the Court, the refusal of the defendant is morally indefensible. [But] to *compel* the Defendant to submit to an intrusion of his body would change every concept and principle upon which our society is founded. To do so would defeat the sanctity of the individual and would impose a rule which would know no limits and one could not imagine where the line would be drawn." *McFall v. Schimp*, cited in Gallagher, "Fetal Personhood," 106.

48. The image here is similar in mood to the one Euripides creates in *The Bacchae* (lines 1200–1308): Agave's moment of recognition and horror when she sees clearly that the "lion" she has hunted and savagely (not to mention ecstatically) dismembered

under the spell of Dionysus was in fact her own son (Chicago: University of Chicago Press, 1958).

49. Maggie Garb, "Abortion Foes Give Birth to a 'Syndrome,' " *In These Times* (February 22–March 1, 1989), 3.

50. Women Exploited by Abortion, "Before You Make the Decision," and "The Pain That Follows: Coping After an Abortion" (pamphlets, no dates).

51. Jill Eisen, "Drawing the Line: Reproductive Technology," cited in Christine Overall, *Ethics and Human Reproduction* (Boston: Allen & Unwin, 1987), 142.

52. See, for example, Peter Roberts, "The Brennen Story: A Small Miracle of Creation," *Test-Tube Babies*, ed. Peter Singer and William Walters (Melbourne: Oxford University Press, 1982), especially 13–16; Isabel Bainbridge, "With Child in Mind: The Experience of a Potential IVF Mother," *Test-Tube Babies*, especially 120–22.

53. *Time*, "The Saddest Epidemic" (September 10, 1984), 50.

54. Cited in Michelle Stanworth, "Reproductive Technologies and the Deconstruction of Motherhood," *Reproductive Technologies: Gender, Motherhood and Medicine*, ed. Michelle Stanworth (Minneapolis: University of Minnesota Press, 1987), 15.

55. The statistic is from Ann Snitow, "The Paradox of Birth Technology: Exploring the Good, the Bad and the Scary," *Ms.* (December 1986), 48.

56. *Life* (June 1987).

57. When particular groups of women (and men) are seen as "breeders," their infertility comes to be regarded as a "solution" rather than a problem.

58. Susan E. Davies, ed., *Women Under Attack: Victories, Backlash and the Fight for Reproductive Freedom* (Pamphlet 7, Committee for Abortion Rights and Against Sterilization Abuse) (Boston: South End, 1988), 31.

59. Carl Wood and Ann Westmore, *Test-Tube Conception* (Englewood Cliffs, N.J.: Prentice-Hall, 1983), 96.

60. *Life* (February 1989), 54–55.

61. *Newsweek* (August 11, 1986).

62. *Life* (February 1989), 55.

63. *New York Times* (April 1, 1987).

64. Worth mentioning here is the ruling on surrogacy issued by the Australian National Health and Medical Research Council. The Council determined that "because of current inability to determine or define motherhood in this context [surrogate motherhood], this situation [was] not yet capable of resolution." Cited in Singer and Wells, *Making Babies*, 97. And about the preferability of ectogenesis or extracorporeal reproduction to surrogacy, Singer and Wells themselves continue: "If, for instance, early experience with surrogacy showed that surrogate *mothers* could not be relied upon to give up *to their genetic parents* the children they had carried, ectogenesis might be thought better than a battle over custody. Evidence that surrogate mothers frequently smoked or took alcohol or drugs that caused harm to the *baby* might be another reason for preferring the strictly controlled artificial environment" (emphasis added), 119.

65. Mary Meehan, cited in Nat Hentoff, "How Can the Left Be Against Life?" *Village Voice* (July 18, 1985), 17.

66. *New York Times* (April 22, 1989).

67. Ibid.

68. *San Francisco Chronicle* (August 11, 1989). The conceptual madness in the analogy is that this woman simultaneously becomes Eichmann and the Allied troops.

Moreover, to fulfill her "rescue mission" all seven "preborn children" must be implanted and brought to term—an undertaking fundamentally at odds with the very rationale of multiple egg harvesting for achieving an IVF pregnancy.

69. *San Francisco Chronicle* (August 11, 1989).

How to Have Theory in an Epidemic:
The Evolution of AIDS Treatment Activism
Paula A. Treichler

> *I got the drug right here, it's called acyclovir,*
> *And though it's used for herpes I have no fear.*
> *Can do, can do, my Doc says the drug can do.*
> *If she says the drug can do, can do, can do.*
> —Ron Goldberg, "Fugue for Drug Trials" [1]

A remarkable development in the evolution of the AIDS epidemic is the crusade of AIDS activists for the testing and release of experimental drugs by the U.S. Food and Drug Administration (FDA) and for participation in the design and implementation of clinical drug trials. The struggle over AIDS drug trials and treatments requires sophisticated technical information about the structure and functions of the FDA and the U.S. Department of Health and Human Services (HHS); about the process and economics of developing, evaluating, and releasing new drugs; about the conceptual and statistical grounds on which standards for clinical drug trials are based; about drugs themselves—where they come from and how they work; about viruses in general and HIV in particular; and about AIDS, HIV infection, and how drugs might act to bring about prevention, retardation, or cure.

The epigraph above is the opening verse of "Fugue for Drug Trials," Ron Goldberg's adaptation of the famous curtain-raiser from *Guys and Dolls* in which each of three seasoned denizens of the New York racetrack, armed with statistics, insider tips, and the daily *Racing Form,* claims, "I got the horse right here." Performed by Goldberg and colleagues in March 1989 after a successful demonstration by New York AIDS activists against Mayor Koch and City Hall, "Fugue for

Drug Trials" captures the Runyonesque combination of unquench-able optimism and bitter experience that characterizes much of AIDS treatment activism. It points to the enormously complex and often agonizing decisions about diagnosis, treatment, and the doctor-patient relationship that must be made daily—individually and collec-tively, personally and politically—by people living with HIV disease. And it suggests the degree of technical sophistication required in the age of AIDS, together with the element of ironic distancing, even of camp, that is sometimes used to frame and contain technological overload.

This essay examines the contributions of AIDS activism to the fight against AIDS and to progressive political action more broadly; it draws on published scholarship and newspaper coverage, discus-sions and interviews with participants, reports and other documents from activist groups, and attendance at conferences and related events. I first review the evolution of drug regulation in the United States and the intervention of AIDS activism in that process. I then look more closely at these interventions, using the debate over the drug zidovudine (an antiviral drug also known as azidothymidine, or AZT) to illuminate the theories and practices that are shaping AIDS research, treatment, and activism. I argue that this debate represents overdetermined cultural narratives about scientific proof, professional identity, the doctor-patient contract, and nature itself. These will con-tinue to shape developments to come. But the debate about AZT, like AIDS treatment activism in general, is inevitably about the uses and consequences of technology and biomedical theory in everyday life. Hence it is a debate, with mortal stakes, about the evolution, value, and possible limits of a radically democratic technoculture.

The Industrial-Regulatory Loop

As the bearer of organic disease, the object of study made manifest, the body is denied its social and cultural embodiment and comes to stand for the disease itself. Hence research on the body becomes legitimated; the body becomes the object of research (a kind of walk-in, skin-encapsulated test tube) and people become the legitimate subjects of research.

—Meurig Horton, "Bugs, Drugs and Placebos"

*Many in Dr. Young's audience held up watches with alarms ringing
to underscore their argument that people with AIDS cannot wait
while drugs are put through years of testing before they are permitted
to be used.*

—Philip Boffey, *New York Times,* July 25, 1988

The public revelations of Nazi experimentation with human subjects during World War II led to the international adoption of the Nuremberg Code. Although the possibility of regulating human experimentation in the United States was discussed by the U.S. Public Health Service (PHS) during the immediate postwar period, medical scientists resisted regulations and guidelines on the grounds that the Nazi experiments were aberrations, carried out by deranged individuals. The thalidomide tragedy, however, led to the passage in 1962 of amendments to the 1906 Pure Food and Drug Act, also known as the Kefauver-Harris Amendments. For the first time, federal law explicitly imposed controls on human experimentation and directed doctors to inform patients when they were being given drugs experimentally.

This history of regulation has created a highly contested arena in which social and ethical concerns for individual rights continually confront the needs and principles of scientific investigation evolved by biomedical researchers. The guidelines of the 1964 World Health Organization (WHO) Declaration of Helsinki, for example, set out in less legalistic terms than the Nuremberg Code the broad ethical principles that should govern research with human subjects. Although the declaration was widely endorsed by medical organizations in the United States, strong objections to actual regulations persisted, and, not surprisingly, a National Institutes of Health (NIH) report that same year revealed no generally accepted codes of clinical research among medical scientists. In response, the PHS issued in 1966 Policy and Procedure Order Number 129, that agency's first set of guidelines on clinical research and training grants. Revised again in 1969 and in 1971, the guidelines challenged the widespread premise that the judgment and integrity of medical researchers provide sufficient protection of the rights and welfare of research subjects. New FDA regulations in 1971 mandated the review of experimental protocols and provisions for obtaining informed consent. Yet these regulations concerned procedural rather than ethical principles; moreover, they did not apply to the federal government's own research.

In 1972, however, the Associated Press broke the story of the Tuskegee Syphilis Study (Jones 1981). The American public learned that since 1932 the PHS had been conducting research in the South on several hundred black men with syphilis, providing them neither information about their condition nor medical treatment. The PHS responded to the public storm — as it had responded to earlier private objections — by arguing that the goal of the research was to learn the natural history of untreated syphilis, that treatment would have compromised this goal, that the men would not have understood their condition or benefited from treatment, and that critics of the research did not understand science.

But this response was no longer good enough; indeed, it was not good at all. After Senator Edward M. Kennedy in 1973 conducted Senate hearings on human experimentation, an ad hoc panel was created to investigate the Tuskegee study. Loud and uniform public outrage persisted, and a class action suit was filed on behalf of the men with syphilis and their families. As a result, tougher controls for clinical research and protections for patients were finally put in place. The National Research Act of 1974 established the National Commission for the Protection of Human Subjects of Biomedical and Behavioral Research. Reflecting the growing cultural consensus that science should take social and ethical concerns into account, the commission was charged with identifying "the basic ethical principles that should underlie the conduct of biomedical and behavioral research" and developing guidelines to ensure that research be conducted in accordance with those principles. Such ethical principles, outlined in the final report of the President's Commission on Ethical Problems in Medicine and Biomedical and Behavioral Research (President's Commission, 1983), include autonomy, beneficence, and justice.[2]

The FDA has thus evolved as a complicated repository of history, incorporating the products of liberal reform, biomedical achievement, and the free market. Charged with protecting the public from dangerous or useless drugs made available by inexpert clinicians or profit-seeking drug companies, the agency stands in an inevitably close relation to clinical medicine and pharmaceutical manufacturing as well as to biomedical research.

The FDA's process for overseeing the development of new drugs embodies this uniquely contradictory historical legacy. Studies of experimental drugs and therapies are called *clinical trials,* defined by Levine and Lebacqz (1979, 728) as "a class of research activities designed to develop generalizable knowledge about the safety and/or efficacy of either validated or nonvalidated practices." Thus designed to determine both safety and efficacy, clinical trials require that a drug be studied first in vitro (in a test tube), then in vivo (in living beings — animals and humans). A plan, or protocol, to study the drug in human beings is submitted by the drug's sponsor to the FDA; it must include results of earlier in vitro and animal testing and describe in detail how it plans to carry out the clinical (human) trial process. The FDA and its Institutional Review Board (IRB) must approve the drug and the protocol before the drug can be released for experimental use on human beings; the IRB is specifically charged with protecting the rights of human subjects and ensuring that informed consent provisions are adequate. The research itself is carried out by the drug's sponsor; what the FDA and IRB see is the application for permission to proceed with research that, with all its supporting documentation, may run to 100,000 pages.

If the FDA decides the drug and the protocol will not place subjects at unreasonable risk, the drug is given Investigational New Drug (IND) status, granting its sponsor the right to begin human testing. Phase I trials, typically lasting less than a year and involving only a small number of subjects, establish the drug's toxicity and the highest dose that can be tolerated (Maximum Tolerated Dose). Phase II trials test the drug under highly controlled conditions in wider populations to determine efficacy (including Minimum Effective Dose) and to provide more information on safety (toxicity and side effects). "Controlled" conditions seek to ensure that any demonstrated effectiveness can legitimately be attributed to the drug being tested: in a standard placebo study, one group is given the drug and a control group is given an identical but innocuous medication; in a double-blind study, neither the experimenter nor the subject knows which is which. Phase III trials seek to confirm the drug's usefulness and to obtain more detailed information about dose, duration, efficacy at various stages and in various populations, and rarer side effects. Most drugs throughout the clinical trial process are tested primarily on white males. Approximately 25 percent of the drugs that enter clinical trials successfully complete Phase III trials, at which point the sponsor may file a New Drug Application. If approved, the drug is placed on the market, where Phase IV studies test its actual performance in the "real world." If the sponsors of a drug (including a vitamin or an herb) claim unsupported therapeutic effects when they market it, or if the drug causes unforeseen problems, it can be recalled by the FDA at this stage.

Overall, the drug approval process is laborious: eight years is the average time it takes for a drug to move from test-tube research to final market distribution. Two alternative regulatory options, however, are available. First, by filing an application for Compassionate Use IND, a physician can request permission from the drug's sponsor and the FDA to administer a drug to a patient who is seriously ill. Second, if tests before controlled clinical trials are completed appear to demonstrate that a drug will not put patients at unreasonable risk, the drug sponsor can apply for Treatment IND status to distribute the drug to patients with immediately life-threatening conditions; continuing research to establish efficacy accompanies the release for treatment. Neither option is as rational as it sounds: applications are time-consuming to produce and often rejected, and, even if approved, typically make a drug available only on a small scale.

Problems inherent in this process are obvious, and they are not helped by the intimate connection between federal regulations and corporate interests that protects the process itself from open scrutiny on the grounds of trade secrecy. AIDS advocates, who seek to get more drugs tested and approved rapidly, denounce both the secrecy and the glacial pace (e.g., the *Treatment and Data Handbook* [Bohne et al. 1989, 6] charges the FDA with "murderous lies and bureaucratic sloth"). They note that the industrial regulatory loop is also a revolving door: in the satiric video *Rockville Is Burning,* the character playing an FDA official advises AIDS advocates that if they want information, they should "talk to Burroughs Wellcome—that's where I'm going when I leave this pissant job—that's where the *real* money is" (Huff and Wave 3 1989). Activists are by no means alone in their view. Critics charge that the federal anti-AIDS effort is underfunded and understaffed (Presidential Commission 1988) and that the drug approval process is "politically contaminated" (quoted in Mahar 1989, 22). Medical ethicist Carole Levine describes the clinical trials process as a system "designed to be cautious. It's designed to stop bad drugs—not to hurry up good drugs" (quoted in Zonona 1988).

One proposed modification for AIDS drugs would eliminate all restrictions on access for patients diagnosed with full-blown AIDS (Krim 1986, 1987). A second proposal takes a laissez-faire approach and recommends that informed consent be the only determinant of access (Gieringer 1987; "An AIDS Crisis Proposal" 1988). While the FDA has not adopted either proposal, it has broadened and liberalized the category of Treatment IND (Young et al. 1988; 22 *Federal Register* no. 99, May 22, 1987, 19467–77). Yet, under Treatment IND drug companies can charge for drugs, not only reversing the standard practice under conventional clinical trials of providing drugs free but also lessening the incentive of drug companies to complete all stages of testing, since they can begin to recover their costs before safety and efficacy have been fully established. Regulations, moreover, cannot fully resolve fundamental conflicts of interest between experimental science and patient care, between the need to obtain sound aggregate data and the duty to protect the rights of individual patients, between the duty to guard the public from unsafe or useless drugs and the need to make potentially useful drugs available rapidly, and between the willingness of afflicted people to risk anything and their special vulnerability to nonvalidated drugs and treatments (see Freedman and McGill Boston Research Group 1989 for a review of these issues with regard to AIDS therapies).

The history I have sketched is studded with these conflicts. While the Tuskegee study clearly abused any reasonable definition of an experimental subject, the format and regulations for clinical drug trials were developed precisely to determine safety and efficacy by stringent scientific criteria. Failure to apply such criteria was what led to the thalidomide tragedy as well as to widespread side effects in women used as subjects to develop oral contraceptives, the drug DES, and the IUD. It was, in fact, these abuses that spurred formation of the women's health movement and eventually induced further regulation: in 1974 the U.S. Department of Health, Education and Welfare developed guidelines to curb sterilization abuse and in 1978 the FDA gained statutory power to regulate medical devices. Yet as Doris Haire's 1984 report to Congress for the National Women's Health Network makes clear, the regulatory process is full of loopholes: FDA approval does not mean the drug has been approved as safe for general use; safety has usually not been tested in nonmale populations, including among the elderly, infants and children, pregnant women, and lactating mothers; no laws or regulations prohibit a doctor from prescribing or administering a nonapproved drug; the FDA often accepts manufacturers' safety data on faith, accepts unpublished data as evidence, may be influenced by economic pressures, does not evaluate drugs in combination, and shields safety research from public scrutiny.[3] During the 1970s, moreover, enthusiasm for tighter oversight waned as the entire health care system began to shift from regulation toward a competitive market model.

The AIDS epidemic came to pass, then, in a health care system fraught with unsolved problems and incompatible philosophical commitments. Add to this the conservative political appointments and budget cuts of the Reagan presidency, a slashed and demoralized staff at the FDA, a bandwagon enthusiasm for deregulation, a renewed faith in the invisible hand of natural market forces, and a growing awareness in Congress of AIDS as a sensitive political issue, and one better understands the terrain from which "AIDS drugs" have so slowly and inconsistently emerged.

Yet, significantly, what are loopholes in one context may be footholds in another. The gaps in the FDA oversight that feminists and consumer advocates criticize are providing space for alternative approaches to AIDS drugs and treatment to take shape. By experimenting with combinations of legal, unregulated, and unapproved drugs and other potentially therapeutic substances, persons with AIDS (PWAs) and their advocates are establishing a body of personal testimony, anecdotal experience, and technological expertise sufficient to challenge the federal effort. They are accomplishing this in three ways: (1) by providing drugs, treatments, and technical expertise through underground or alternative channels; (2) by working cooperatively with selected scientists, physicians, and drug manufacturers to explore new therapies and develop alternative strategies for testing and distributing new drugs and treatments; and (3) by directly challenging the FDA and related scientific agencies involved in federal AIDS efforts.

First, as physicians and patients became increasingly convinced that treatment was an important variable in disease progression, guerrilla clinics and buyers' clubs formed to provide unapproved or illegal drugs and treatments. Such organizations as Project Inform in San Francisco and such newsletters as *Treatment Issues, The Body Positive, Positively Healthy*, and *AIDS Treatment News* were founded

Treatment options. Still from video *Work Your Body* (Bordowitz and Carlomusto 1988).

to report and analyze the development of validated and nonvalidated therapies, scientific developments, and FDA actions and inactions (see Callen 1987). Project Inform and the AIDS Coalition to Unleash Power (ACT UP) used the Freedom of Information Act (FOIA) to obtain documents pertaining to the FDA's approval of the drug AZT and proceeded to analyze flaws in the process and in the drug (see, e.g., Sonnabend 1989; Lauritsen 1989; see also Erni forthcoming). Some medical researchers and clinicians, accordingly, began to argue that people with AIDS were too well informed and technologically expert to submit passively to "controlled" clinical trials. Subjects enrolled in placebo trials, for example, would routinely take their samples to be analyzed privately; those on the "real" drug would then divide their supply with those on placebos. Others refused to enroll in clinical trials at all on the grounds that placebo studies for people with a deadly disease were unethical. Scientists familiar with these practices argued that since PWAs were inevitably going to experiment with diverse treatment possibilities, scientists should find ways to study the effects systematically.

A I D S
TREATMENT
N E W S

Issue Number 78
May 5, 1989

Published biweekly by
John S. James
P.O. Box 411256
San Francisco, CA 94141
415/ 255-0588

Contents

Compound Q Warning, and Update

Compound Q, an experimental AIDS treatment extracted from the root tuber of a Chinese cucumber, has received wide publicity in the last month. On May 5 we heard the first report of a severe adverse reaction to a bogus "compound Q", apparently homemade from the root which was obtained from a health-food store, and injected. According to Martin Delaney of Project Inform, who is now warning buyers' clubs, the person almost died as a result, and was in intensive care for three days. This case occurred in Kansas City.

We have also heard that some health-food stores are exploiting the situation and promoting a dried root or extract by suggesting that it contains compound Q. People should know (1) that the root also contains lectins, which are poisonous when injected because they cause blood cells to clump together, which can cause heart attacks or strokes, and (2) that compound Q (which is a protein called trichosanthin) is almost certainly destroyed by drying, so the dried root used as an herbal medicine for other purposes does not contain the active ingredient.

It is generally believed that a good-quality equivalent of compound Q does exist in China, and has been used there for other purposes for several years (see *AIDS Treatment News* #77, pages 1-2 and page 5). However, this drug is tightly controlled and very difficult to obtain. We have heard from knowledgeable persons (but have not yet been able to confirm independently) that only half a million doses a year are manufactured, all by one factory in or near Shanghai, and that some of it did reach a few persons with AIDS in the U.S. While extracting the active ingredient (trichosanthin) from the Chinese cucumber root is not too difficult for a protein chemist, there are practical problems, especially the need to obtain large quantities of the fresh or frozen root, as well as the usual difficulties of setting up effective manufacturing and quality control for pharmaceuticals.

Any credible, good-quality data which may develop from use of the Chinese compound-Q equivalent would be very important in speeding the authorized clinical trials. At this time, the only clinical trial planned anywhere in the world is a "phase I" study to take place at San Francisco General Hospital. This trial may be slowed by the current budget crisis of the City and County of San Francisco, since hospitalization is required for the study but there is not enough funding to staff the nursing support for the hospital beds.

The San Francisco trial will also be slow because it is designed primarily to test for toxicity and determine the maximum tolerated dose, not to determine whether the drug can help patients. A tiny dose which no one believes could be effective will be tried first, followed by a wait to

Opening page of *AIDS Treatment News,* May 5, 1989.

Second, in the absence of federal coordination and treatment guidelines, private physicians were prescribing drugs and treatments supported by their experience and that of their patients and colleagues. A notable discovery was the effectiveness of aggressive treatment of the symptoms and opportunistic infections associated with

HIV infection; in particular, a regimen that included regular prophylaxis for pneumocystis carinii pneumonia (PCP)—the most common
opportunistic infection in people with AIDS, which about 85 percent
experience at some point in their illness—was found by many physicians to promote health and to retard disease progression significantly (Altman 1988). Yet in the absence of official scientific and clinical recommendations, early intervention in general and specific
treatment regimens in particular had limited ability to save lives outside the cities where such forms of treatment were standard practice
(Callen 1989). By this point important relationships had been formed
among physicians, medical researchers, and patients over the question of AIDS treatments. Many of these physicians had large practices
of AIDS patients and were frustrated too. Further, many understood
the drug approval process and could serve as informants and con-
sultants. And their patients were ready to try new drugs, a fact of interest to pharmaceutical companies unable to get drugs tested. Congressional committees were exploring ways to facilitate these efforts.
The desire to "fast track" drugs also attracted supporters of deregulation and potential investors in AIDS drug development. These coalitions formed a potent lobby, generating proposals, as early as 1986,
for community-based AIDS treatment research programs to speed
testing and release of safe drugs for treatment.[4]

Third, political activism directed at the existing system sought to
modify most features of the FDA oversight process described above.
Congressional hearings were held in 1987 as a result of the agitation;
as noted above, the FDA subsequently approved Treatment IND status for several AIDS drugs. By 1988, enough knowledge of the drug
development and approval process had been amassed so that FDA
Commissioner Frank Young could be authoritatively challenged
when he stated in the summer of 1988 that only two new drugs could
be approved before 1991. Following a day-long protest by AIDS activists at the FDA in October 1988, which received extensive media
coverage, the FDA relaxed some regulations; held discussions with
AIDS advocates, community physicians, and others; and approved
several AIDS drugs targeted by advocates. Notable among these was
aerosolized pentamidine, a drug used in aerosol form to prevent and
treat PCP; a lifesaving treatment for people who cannot tolerate more
conventional PCP drugs (including Bactrim, Dapsone, and injected
pentamidine) and the target of sustained activism, the drug was at

A laboratory record showing interferon blood levels of a
patient being treated at a major medical center with the
Roche prescription medicine Roferon®-A.

Press kits from pharmaceutical companies create photo opportunities for new
drugs. Still from Hoffman-La Roche press kit, June 1989.

last given official sanction as an experimental drug (Torres 1989). At
about the same time, standardized guidelines for treating PCP were
released by the Centers for Disease Control (1989).

In July 1988, the FDA reversed an earlier stand and approved the
import of unapproved AIDS drugs (though in small quantities and
for personal use only; Boffey 1988c). Activist protests, and the media
coverage of them, influenced the makers of approved drugs to lower
their costs. At the Fifth International Conference on AIDS in Montreal
in June 1989, ACT UP presented its *National AIDS Treatment Research Agenda,* a comprehensive plan for AIDS treatment research.
Extensive discussions of the agenda with FDA officials followed, focusing particularly on the need to expand access to clinical trials and
better meet the ethical principles of autonomy, beneficence, and justice. Out of these discussions came more formal proposals to test
promising new drugs through community-based research and make
them accessible through parallel release programs (Harrington 1989,
1990). Federal and other funding was made available for such community-based treatment research in 1989. And when Project Inform

atic patients; at the 30 million IU dose, 47 percent of asymptomatic patients responded versus 9 percent of symptomatic patients.

Patients who responded to treatment lived significantly longer than nonresponders (median survival 23 months versus 10 months). Intron-A was generally well tolerated, although most patients experienced flu-like symptoms, including fever, chills, muscle pain, and fatigue. In these studies, only six patients (six percent) discontinued treatment because of adverse reactions.

The clinical responses produced by all three regimens confirm the previously reported activity of alpha interferon against AIDS-related KS.[2,10] Intron-A was approved for subcutaneous or intramuscular injection three times per week and can be administered as outpatient therapy. In other studies, Intron-A has

chloride, five milligrams per liter. However, the presence of the antibiotic is not detectable in the final product. Roferon-A is supplied as an injectable solution or as a sterile powder for injection with its accompanying diluent.

Six studies were conducted to determine the effect of Roferon-A on AIDS-related KS. Three-hundred-and-fifty patients were given a total of 3 to 54 million IU daily. Four dosage regimens of Roferon-A were evaluated for initial induction. Thirty-nine patients received three million IU daily; 99 patients received an escalating regimen of 3 million, 9 million,

Roferon®-A is manufactured by Hoffmann-La Roche, Inc., Nutley, NJ.

higher than 36 million IU daily were associated with unacceptable toxicity.

Dosage Administration

Indicator lesion measurements and total lesion count should be performed before initiation of therapy. These parameters should be monitored periodi-

Press kit photo reproduced in *AIDS Patient Care,* June 1989 (p. 19).

initiated unauthorized testing of trichosanthin (Compound Q) in 1989, the FDA advised against the trials but did not forbid them; when official trials were initiated, the FDA agreed to a protocol negotiated by Project Inform that would incorporate the patients from the underground trials (*Treatment Issues,* 3 [May 15, 1989], 1–2).

This challenge to the federal regulatory process reflects a significant understanding of the health care system; potentially, a broad range of policies for all drugs will be affected. AIDS treatment activism, including proposals for parallel release and community-based AIDS treatment research as well as continuing interest in a broad range of nonvalidated therapies, is tied to the AIDS movement's evaluation of technology and its determination to make technological resources available to people living with AIDS. This goal is in many ways incompatible with existing scientific and medical practice, with the current capabilities of an overburdened health care system, and with a long-standing radical distrust of technology. Also at stake, then, is the potential for the growth of a radically democratic technoculture. By this I mean that the strongest challenge to current conditions comes not from those who dismiss or denounce technology, but from those who seek to seize it for progressive political purposes and for the deployment of science and scientific theory in everyday life.

Technology and Resistance

In the beginning, those people had a blanket disgust with us. And it was mutual. Scientists said all trials should be restricted, rigid and slow. The gay groups said we were killing people with red tape. When the smoke cleared we realized that much of their criticism was absolutely valid.
                                    —Anthony S. Fauci, *Washington Post,* July 2, 1989

People tell me, "You've lost your resistance." I say, "Not yet!"
                                    —Herbert Daniel, *Life Before Death*

You could think of AIDS treatment activism as a postmodern post-Stonewall reworking of *Walden, Our Bodies Ourselves,* the *New England Journal of Medicine,* and *The Scarlet Pimpernel.* Its mix of strategies and sensibilities is evident in *Rockville Is Burning* (Huff and Wave 3 1989), a video-theater piece that takes the October 1988 AIDS action at the FDA as its starting point. The video opens with a Dan Ratheresque network anchor reporting the takeover and burning of the FDA by "AIDS terrorists," a "shadowy group that calls itself the New Center for Drugs and Biologics" and is described by inside sources as "extremely well informed and extremely dangerous." Then, during a commercial break ("When we come back: more on that puppy trapped in a well shaft in Texas"), three "terrorists"—two men and one woman—suddenly take over the broadcast studio in order to give their own account of AIDS treatment research. Using live "uplinks" from around the country to support their charges and illustrate their demands, they lay out a basic critique of the FDA's process (I return to this below). They conclude their broadcast with the following speech, which they deliver sequentially:

> Two years ago most of us never would have conceived of marching in the streets, much less using flamethrowers or hijacking television sets. The truth is we never saw what was happening around us. We never saw beyond the facade.
>
> That is, until it hit home. Until we realized that the system was killing us and we started trying to figure out what was happening.

> Slowly we educated ourselves. We began to analyze the
> bureaucracy and the politics. We read stacks of tedious protocols
> and contracts. We learned medical terminology and the tricks of the
> budget manipulators.
>   And slowly a pattern began to emerge. The very people with the
> firsthand knowledge of the epidemic were the last to be consulted.
> And while we were waiting for kind words and crumbs from the
> liberal managers of the epidemic we realized that they were simply
> links in the chain of command. It wasn't a question of saving lives
> or even of saving money—it was about power.
>   But when the first PWA chose to sit down and be dragged off in
> the middle of Wall Street, we started to take back some of that
> power.

This version of AIDS treatment activism, probably best exemplified in real life by ACT UP, invokes several essential elements of the movement: a vision of the power structure that calls for unleashing the power and knowledge of resistant forces; expertise about technology and science, the politics of the federal bureaucracy, biomedical research, and economics; self-education; and the use of tactics including civil disobedience, lawbreaking, infiltration, and seizing control of the media.

These strategies are not entirely recent.[5] Almost from the beginning of the AIDS epidemic in the United States, gay men have attempted, individually and collectively, to conceptualize scientific and clinical explanations of acquired immune deficiency, to articulate the meaning of the epidemic, and to decide for themselves what to do about it. The immediate historical context for this grass-roots approach to the AIDS crisis includes the antiauthoritarian legacy of post-Stonewall gay liberation; the successful struggle within the American Psychiatric Association by gay psychiatrists and gay rights groups in the early 1970s to remove homosexuality as an official category of mental disorder (Bayer 1981); the celebrated collaboration among physicians, research scientists, and the gay community in the clinical trials of a Hepatitis B vaccine (Goodfield 1985); and the philosophy, knowledge, and tactics developed by the women's health movement. Within this context, education and prevention efforts began in the gay community even before there was general acknowledgment of a fatal epidemic disease. Randy Shilts (1987, 108) documents the élan with which Bobbi Campbell carried out his campaign as "the KS poster boy" in San Francisco's Castro District, and Frances

FitzGerald (1986) details the lengthy debates within the gay community about what to do.

While education efforts by health professionals advised abstention from "promiscuity" in general and specific "high-risk" sexual practices in particular, some gay men approached the crisis as a technical problem. The pamphlet *How to Have Sex in an Epidemic* (Callen and Berkowitz 1983) did not try to persuade gay men to abstain from sex or to relinquish the pursuit of sexual pleasure to atone for the excesses of the 1970s; instead, it provided an analysis that grew out of, rather than retreated from, the gay liberation movement. It analyzed the body (and specifically the gay male body) as an environment to be respected, technically manipulated, and cultivated to foster health rather than disease. For the authors of *How to Have Sex,* published before AIDS was officially linked to a transmissible viral agent, the epidemic was a tale of neither conservative morality nor medical mortality but a crisis in which a unique body of knowledge would be needed. As Douglas Crimp (1988) argues in "How to Have Promiscuity in an Epidemic," titled to celebrate the politics of AIDS activists, the sexual adventures of the 1970s should be seen as a key behavioral resource for inventing new, safer ways to have sex in an epidemic.

Safer sex guidelines, soon taken up by health professionals and other constituencies at risk, are widely acknowledged to have affected both the scope and the public perception of the epidemic. Treatment, in comparison, seemed a will-o'-the-wisp in light of the apparent reality that AIDS was "untreatable" and "incurable." On a local level, AIDS workers and PWAs sometimes discouraged attention to it as a diversion of resources from human services and everyday clinical management, and sometimes, too, as a denial of death (see Douglas and Pinsky 1989 for a fuller analysis). A political effort for basic and treatment research at the federal level continued, producing appropriations from Congress in 1986 for scientific research and drug development (see Shilts 1987 for an account of this struggle).

At the same time, something else was happening at the grass roots. Although prevention was not a high-tech solution, it was nevertheless based on a body of technical knowledge and behavioral experience. By the mid-1980s, communication about living with AIDS through anecdotal reports and newsletters gradually put into circulation the news that more people seemed to be living longer. Some of the reported self-treatments relied on the established literature of holistic medicine and self-care; some testified to ways of clinically managing the opportunistic infections that were the main cause of suffering and death (I have mentioned prophylaxis to prevent and treat PCP); and some described drugs and treatments identified through original research, expert consultation, cross-cultural communication, and/or personal experimentation. These reports testified to the resourcefulness and determination of people who have decided they will try anything, break any law, and do whatever is necessary to get treatments they perceive as needed. Underground networks were set up to provide drugs and treatment: the first guerrilla clinic was founded in San Diego in late 1985 (Geitner 1988); buyers' clubs also formed to obtain and distribute both gray-market (unapproved) and black-market (illegal) drugs, in some cases importing or smuggling drugs in bulk into the United States from Japan, Germany, Mexico, France, or wherever else they were available (Kolata 1988a; Greyson 1989).

An important catalyst for directing these efforts toward a systematic challenge of the industrial-regulatory loop was Larry Kramer's June 1987 speech to the annual Gay and Lesbian Town Meeting in Boston. I report this in some detail because it exemplifies a vision of the federal AIDS effort that helped establish the direction and operating mode of ACT UP and articulated a general shift in emphasis among some activists from prevention to treatment. In this speech, Kramer, a writer and founder of Gay Men's Health Crisis in 1982 and ACT UP in 1987, tells the gay community that their successful efforts to get research funded have been for nothing, because the system isn't working.[6] In seven years, the only drug that's been produced is AZT, he says, which is highly toxic. Gay men must have a death wish to sit back and let themselves be killed, and he lists all the things they could be doing but aren't to make the federal system do its job.

Kramer's polemic is a brilliant call to activism and one of the most incisive and useful analyses of a government bureaucracy ever written. To emphasize his main thesis, Kramer (1987, 37) dismantles, piece by piece, every institution in which his audience may still have faith:

> No one is in charge of this pandemic, either in this city or this state or this country. It is as simple as that. And certainly no one who is compassionate and understanding and knowledgeable and efficient

is even anywhere near the top of those who are in charge. Almost every person connected with running the AIDS show everywhere is second-rate. I have never come across a bigger assortment of the second-rate in my life. And you have silently and trustingly put your lives in their hands. You—who are first-rate—are silent. And we are going to die for that silence.

The money appropriated by Congress to fight AIDS, says Kramer, is not being spent; for example, the National Institutes of Health was given $47 million just to test new drugs, and they aren't doing it:

When I found out about three months ago that $47 million was actually lying around not being used, when I knew personally that at least a dozen drugs and treatments just as promising as AZT, and in many cases much less toxic, were not being tested and were not legally available to us, I got in my car and drove down to Washington. I wanted to find out what was going on.

Kramer then tells his audience about each federal official in the AIDS chain of command. He says Dr. C. Everett Koop, the U.S. surgeon general, is out of the loop—outspoken but powerless; he says Dr. Otis Bowen, secretary of health and human services and the one official who reports directly to President Ronald Reagan, has still—after seven years—not said anything significant about the AIDS epidemic; he says Dr. Robert Windom, Bowen's assistant, is exceptionally ill informed about AIDS and exceptionally dumb—"If his IQ were any lower," one aide told Kramer, "you'd have to water him" (p. 38); he says Dr. Lowell Harmison, Windom's assistant, actually believes gay people will *intentionally* give blood to pollute the nation's blood supply.

Dr. Harmison reports to Dr. Windom, who reports to Dr. Bowen, who reports to the President. . . .
   I am here to tell you that I know more about AIDS than any of these four inhumane men, and that any one of you here who has AIDS or who tends to someone who has AIDS, or who reads all the newspapers and watches TV, knows more about AIDS than any of these four monsters. And they are the four fuckers who are in charge of AIDS for your government—the bureaucrats who have the ultimate control over your life. (p. 38)

Then he gets to the NIH, directed by Dr. James Wyngaarden, and to Dr. Anthony Fauci, director of the National Institute of Allergies and Infectious Diseases (NIAID) of the NIH,

the single most important name in AIDS today ... who has
probably more effect on your future than anybody else in the
world. ... Dr. Fauci is an ambitious bureaucrat who is the recipient
of all the buck passing and dumping-on from all of the above. He
staggers, without complaint, under his heavy load. No loudmouth
Dr. Koop he. ... Dr. Fauci, of all the names in this article, is
certainly not the enemy. Because he is not, and because I think he
does care, I am even more angry at him for what he is not
doing—no matter what his excuses, and he has many. Instead of
screaming and yelling for help as loud as he can, he tries to make
do, to make nice, to negotiate quietly, to assuage. An ambitious
bureaucrat doesn't make waves.

Yes, Dr. Fauci reports to Dr. Wyngaarden, who reports to Dr.
Windom, who reports to Dr. Bowen, who reports to the President.
(p. 39)

Kramer's house-that-Jack-built exposé of the federal AIDS effort
confirmed in blunt language what official reports were saying more
guardedly. It served as a bulletin from the front, and its content and
anger also suggested a strategy for action. Between June 1987 and
June 1989, ACT UP chapters formed in many U.S. cities with the aim
to get "drugs into bodies" by whatever means possible. In July 1987,
ACT UP New York staged a round-the-clock vigil at Memorial Sloan-
Kettering Hospital in New York, a designated AIDS Treatment Evaluation Unit (ATEU), one of nineteen centers across the country established by the NIH to test new AIDS drugs; with $1.2 million in
funding, Sloan-Kettering had by July 1987 enrolled only thirty-one
patients in drug trials. ACT UP's public protest and factual leafleting
were appreciated by frustrated health care professionals; investigations in the wake of the vigil identified the many points at which the
ATEU system was not working and initiated changes for improving it
(Crimp and Rolston 1990). Calling for a "Manhattan Project" on AIDS,
activists repeatedly used military metaphors to describe their situation. "Living with AIDS," said Vito Russo (1988, 65) at an HHS rally in
October 1988, "is living through a war which is happening only for
those people who are in the trenches." "You cannot underestimate
the therapeutic value of feeling like a soldier in the war against
AIDS," says Dr. Nathaniel Pier (quoted in Zonona 1988). Invoking the
grounds on which this war was being waged, Nancy Wechsler (1988)
wrote of the October 11, 1988, action by twelve hundred demonstra-

tors at the FDA that "by the end of the nine-hour blockade, 176 of us had been arrested, and by current estimates, 18 more Americans had died of AIDS."

If the fight for alternative AIDS treatments within PWA and HIV-infected networks underscores the self-empowerment and antiauthoritarian stance of AIDS activists, the organized challenge underscores a sophisticated understanding of the medical-industrial complex and how to turn its own tactics against it. "ACT UP! FIGHT BACK! FIGHT AIDS!" became ACT UP's working policy, a policy that paired extensive background research with an increasingly professional campaign to educate the media and, in turn, to influence public perception of the treatment crisis. By the Fifth International AIDS Conference in Montreal in June 1989, ACT UP chapters had held successful actions at the Brooklyn Bridge, the Department of Health and Human Services, the FDA, Wall Street, the New York Stock Exchange, *Cosmopolitan* magazine, New York City Hall, Kowa Pharmaceuticals, the Democratic and Republican conventions, the Golden Gate Bridge, Rockefeller Center, the New York state capitol at Albany, University Hospital in Newark, New Jersey, Memorial Sloan-Kettering Hospital, the Los Angeles County Hospital, Shea Stadium, the Hall of Justice in Washington, and meetings of the Presidential AIDS Commission.

What are the goals of AIDS treatment activism? They are tailored quite precisely to the working procedures and principles of the FDA, outlined above. The *National AIDS Treatment Research Agenda* constructed by ACT UP (1989), in consultation with many other groups and projects, calls for changes in basic principles in the testing of AIDS drugs, proposes alternative models for clinical trials, and lists concrete research priorities (drugs and treatments).[7] The principles section of the document calls for greater participation in the design and execution of clinical trials by people with AIDS, people with HIV, and their advocates; rapid testing and distribution of all promising drugs; search for drugs that fight the entire spectrum of HIV's clinical manifestations, not just flashy antiviral drugs; inclusion in clinical trials of women, people of color, children, and others traditionally excluded (including those found to be intolerant to AZT); design of trials for the "real world" of health care in which treatment for infections is given but placebos are not; reasonable inclusion criteria; humane and compassionate evaluations of efficacy; and access to trials and promising treatments regardless of personal income. In addition, the agenda calls for increased funding of the entire drug research and development network and for the establishment of an international up-to-date registry of clinical trials and treatments for HIV infection and related opportunistic infections.

Several pioneering projects outside the federal establishment are now attempting to institute some of these changes. Parallel-track programs would enable the testing of drugs on patients who have been excluded, for a variety of reasons, from conventional experimental trials (Harrington 1989). Community-based AIDS treatment research organizations, such as Project Inform in San Francisco and the Community Research Initiative (CRI) in New York City, work directly with pharmaceutical companies to test drugs outside the rigidly controlled environment of major medical centers. Among other things, Project Inform organized underground trials of Compound Q. The CRI has organized a number of trials, several sponsored by drug companies. The largest trial to date, involving 225 patients, collected data that speeded the approval of aerosolized pentamidine. Sixty physicians participated; the CRI's sponsorship guaranteed the speedy recruitment of research subjects. The CRI also launches quick studies to monitor the effects of the unproven underground treatments used by many AIDS patients. "If people are taking it, that's almost reason enough to study it," says Tom Hannan. "If something works, great. If it is ineffective or harmful, we want to get the word out" (quoted in Zonona 1988).

Some academic researchers question the value of research data gathered by community physicians, and charge that such programs inevitably threaten the integrity of the clinical trials process and destroy the federal AIDS effort. Health care consumer advocates who have fought fiercely for tougher regulation agree (Kolata 1988a). Yet scientific advisory committees of community-based AIDS research programs typically include representatives of academic medicine and basic science research. Both resistance and support reflect the growing ferment over the bureaucratic federal approval process as well as the urgently felt need to get more people with AIDS and HIV infection into clinical drug trials. Indeed, the federal government has now established a $6 million program to promote community-based AIDS treatment research throughout the country; a major goal is to enroll groups traditionally underrepresented in academic clinical

"A shadowy group [of AIDS terrorists] that calls itself the New Center for Drugs and Biologics" takes over a national news broadcast in *Rockville Is Burning* (Huff and Wave 3 1989). Still from color video.

research — HIV-infected women, minorities, drug users, children and infants, and those excluded from other trials for other reasons. Likewise, the American Foundation for AIDS Research announced it would contribute $1 million to fund pilot programs. In part the change is a pragmatic one: if a single "magic bullet" will not emerge to cure AIDS and its various manifestations and complications, the search must be for a combination of agents to keep the virus inactive, to revitalize the immune system, and to treat the range of clinical problems HIV can cause. Thus dozens of drugs must be tested on tens of thousands of volunteer subjects. A serious impediment to this goal has been the failure to enroll enough subjects in controlled clinical trials. Breaking new scientific and ethical ground, these community-based programs seek to produce scientifically valid findings under more flexible conditions than conventional clinical trials and to reconcile the conflicting goals of scientific investigators and human experimental subjects.

The activist monologue that concludes *Rockville Is Burning* efficiently recapitulates the history and principles of AIDS treatment activism, including its commitments to civil disobedience, self-empowerment, technological expertise ("Treatment: Understand it in order to demand it"—Douglas 1989, 2), and action outside the law. *Rockville Is Burning*—like *Seize Control of the FDA,* a documentary video of the FDA action (Bordowitz and Carlomusto 1989), and *The Pink Pimpernel,* a romantic adventure about the smuggling of underground AIDS drugs (Greyson 1989)—demonstrates the manipulation of conventional cultural narratives and representations to tell an alternate story. Finally, AIDS treatment activism does not depend on an us/them division in which the category *us* is good, pure, natural, and human while the category *them* is bad, profit-seeking, contaminated, and cold-bloodedly technological. Rather, it has assembled out of available resources a complex conception of the body and a multilayered strategy for restoring it to health. This conception is framed as provisional, but nevertheless as a theory for everyday life that can be used to guide practical actions. Experiments such as community research initiatives promise to make a unique contribution to the process of producing knowledge. The strength of their guiding theoretical frame lies not in a resistance to orthodox science but in strategic conceptions of "scientific truth" that leave room for action in the face of contradictions. This makes it possible to seek local, partial solutions, and to give more attention to difference and diversity.

AZT on the Head of a Pin

> *I'm afraid that the AZT argument is a kind of magnet for people's anti-establishment feelings. Whereas that might be okay under a lot of circumstances, it's not okay when we're talking about life and death choices. This is a medical decision, not a political or philosophical one.*
>
> —Martin Delaney, 1988

> *Essential Oils are wrung—*
> *The Attar from the Rose*
> *Is not expressed by Suns—alone—*
> *It is the gift of Screws—*
>
> —Emily Dickinson, 675 (H249) "[Essential Oils Are Wrung]"[8]

Lest this sound too utopian, it is important to emphasize that deep differences over theory and practice require continuous negotiation within the networks of those concerned with treatment. As alternatives increasingly become available, and yield failures as well as successes, tensions, disputes, and schisms are inevitable. Indeed, such tensions and disputes already exist—for example, in ongoing disagreements about zidovudine, or AZT (also known as azidothymidine, its pharmaceutical name, and Retrovir, its brand name). As consensus and routinization around treatment with zidovudine evolve, the divisions at the heart of current struggles will be less visible. The AZT debate, then, helps illuminate questions of theory and practice that will continue to put AIDS treatment activism to difficult tests.

Manufactured by the Burroughs-Wellcome Pharmaceutical Company, zidovudine is the only approved drug for the treatment of the spectrum of AIDS-associated problems. Derived from the sperm of herring, it was tried as a cancer chemotherapy 30 years ago and abandoned as too toxic and expensive, then retrieved in the FDA's move to test drugs "off the shelf" for effectiveness against AIDS. As soon as benefits were claimed in corporate-sponsored Phase II trials in September 1986, the placebo trials were cut off and the drug was distributed in limited quantities under a Treatment IND until its full FDA approval in March 1987. In August 1989, NIAID reported that AZT had been found to be beneficial in asymptomatic HIV-positive persons with fewer than 500 T4 cells. According to the published study (Volberding et al. 1990), equal numbers of patients with T4 cell counts below and above 500 were divided according to three conditions: placebo, 500 milligrams of zidovudine daily, or 1,500 milligrams zidovudine daily. Over the year in which the patients were followed, twice as many placebo patients developed AIDS or AIDS-related symptoms as those taking the drug; only 3 percent on the lower dose developed lowered counts of red and white blood cells, in contrast with 12 percent on the higher dose; and no statistical difference in efficacy was found between the higher and the lower dose, the latter being less toxic and cheaper. These findings offer a reasonable basis for optimism. The FDA has recently changed the labeling instructions for prescribing, a change that should facilitate reimbursement by insurance or Medicaid. The price of the drug has also changed over time. Because AZT was developed under the Orphan Drug Law, Burroughs-Wellcome has a seven-year monopoly (followed by a seventeen-year use patent). Protests over the cost of the drug ($8,000-$10,000 a year per patient) included pasting "AIDS Profiteer" stickers

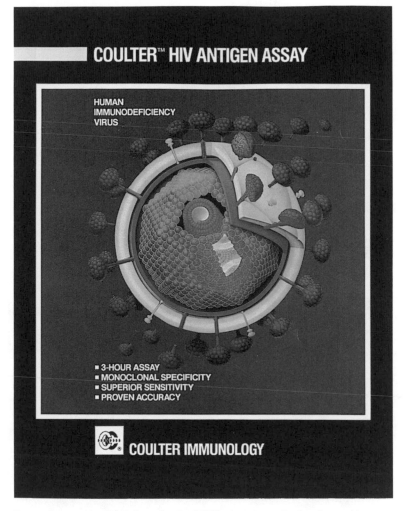

Cover of promotional brochure for the HIV Antigen Assay, Coulter Immunology, Hialeah, Florida, 1989.

on Burroughs-Wellcome products in stores, draping a banner reading "SELL WELLCOME" above the floor of the New York Stock Exchange, and taking over an office at the company's headquarters. Though these activities, together with lobbying by a broad coalition of activists and legislators, forced Burroughs-Wellcome to lower the

AIDS Profiteer stickers, 1989 (ACT UP Outreach Committee). Offset lithography on stick-on Avery labels.

cost of the drug by 20 percent, the company has still not shared its production costs with congressional oversight committees (*Treatment Issues,* 3 [October 30, 1989], 10).

Both physicians and patients in New York have been perceived as uniquely recalcitrant in their resistance to AZT, with critics of the drug particularly opposing its use to prevent disease progression in asymptomatic HIV-infected people. They argue that not only is the drug toxic, it may destroy the very resources that the body needs to resist the destruction of the immune system. The report on AZT that led to its initial approval (Fischl et al. 1987) was widely criticized, yet the drug's supporters argued that its potential benefits justified its release; subsequent studies appear to confirm the positive results of zidovudine, at least over the short run (Friedland 1990).

But though the AZT controversy is gradually being resolved by what is perceived as the accumulation of scientific evidence and clinical experience, it has not been the first such argument, nor will it be the last. Like other debates, it does not represent simply a local disagreement over "facts" and "truth"; it is also the distillation of deep-seated cultural discourses about how facts are produced and truth arrived at and about what values should shape this process. It is instructive, therefore, to examine these overdetermined cultural narratives in action at the second annual conference on AIDS/HIV treatment organized by the Columbia Gay Health Advocacy Project. Held at Columbia University on November 19, 1988, this is described in the program as a conference designed "for the lay person on treatments and health maintenance strategies for people with AIDS, ARC, and asymptomatic HIV infection, featuring a distinguished panel of researchers, clinicians, and activists."

Panelists for the conference were told in advance to expect a well-informed and technically knowledgeable audience. On the morning of the conference, organizer Laura Pinsky cautioned the audience that the day would be long and feelings would run high. Therefore:

> We want to ask for a lot of cooperation in terms of not booing and hissing, which is going to take up time, alienate the panelists, and make it hard for us to invite people back next year (not to mention scaring the people from California). (in Douglas 1989, 1)

The opening speaker in the session on zidovudine was Craig Metroka, an M.D.-Ph.D. at St. Luke's/Roosevelt Hospital in New York and an editor of the *AIDS Targeted Information Newsletter*. Metroka first outlined the life cycle of HIV, noting six points in its replication process at which antiviral drugs can potentially intervene. AZT is one of a group of drugs called *nucleoside analogs* that interrupt replication fairly early on, binding with the virus's reverse transcriptase (its transcribing mechanism) and preventing it from copying its DNA into that of the host cell. Representing academic research tempered by clinical experience, Metroka is involved in the real world of AIDS treatment but is also equipped to analyze technical data in some detail. His presentation summarized what studies appeared to show as of November 1988. The results of the Phase II study showed improved short-term survival, reduced frequency and severity of opportunistic infections, and delayed progression to AIDS; it encouraged weight gain and improved overall functioning. Its toxicity was in

Replication

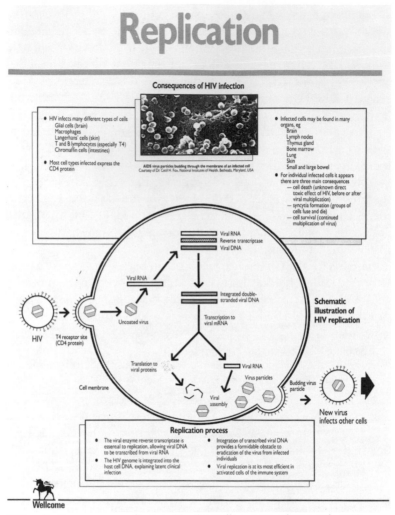

HIV replication process, as illustrated by the Wellcome Foundation, Ltd., 1989.

some cases substantial, including nausea, muscle ache, headache, fever, skin rash, and dementia. Because it interrupts cell replication, especially in bone marrow, it can severely deplete red and white blood cells, causing fatigue, shortness of breath, and severe anemia; after eight months, some patients experienced severe leg pain (from muscle wasting).

Metroka then addressed the most controversial question: when to start AZT treatment. We know most about the most severe cases, he argued; much less is known about asymptomatic seropositive patients. While a European study of 300 patients had shown that the usefulness of AZT seems to decline after six months, the results of large studies were not expected until 1991 (but see Volberding et al. 1990—these studies were released early). High toxicity must be taken into account, thus Metroka's bottom line as of November 1988 was to "use AZT only in those groups in which a survival benefit has been claimed, that is, in patients with fewer than 200 T4 cells" (in Douglas 1989, 25). (AZT is now routinely prescribed for patients whose T4 cell count is under 500.) He makes exceptions for people whose T4 cell count is not so low if other clinical and laboratory findings suggest that their health is deteriorating. Metroka is conservative in wanting to see published scientific research before reaching conclusions about treatment, yet more flexible than some biomedical scientists would be in his willingness to depart from strictly orthodox protocols. Accordingly, he provides PCP prophylaxis for patients with low T4 cell counts and argues for the discontinuation of placebo testing and the concurrent provision of helpful drugs.

Martin Delaney, founder of Project Inform in San Francisco and a central figure in negotiations between the FDA and the PWA community, then described the West Coast experience with AZT, which has been somewhat different. An early organizer of underground drug runs, Delaney operates according to fairly pragmatic rules and is loyal to a real-world constituency, not to a body of abstract biomedical principles. Speaking at the conference, he confirmed the negative side effects, agreed that the drug is overpriced, that Burroughs-Wellcome is "ripping us off," and that the drug produces serious toxicities. But, despite flaws in the original studies (Delaney was one of those who obtained the FDA data through the FOIA), studies taken together now show, he believes, a clear pattern of usefulness and benefits that have not been duplicated by any other drug. Delaney commented as follows on coastal differences:

> AZT use has not become as much of a religious debate on the West Coast as it has on the East Coast, for a variety of reasons. But I think it's important in entering this discussion to realize that AZT is not the enemy, and the people who disagree over AZT use are not the enemy. AIDS is the enemy, and we are all seeking to find solutions. (in Douglas 1989, 25)

Delaney then went on to outline new thinking about AZT. As treatment has matured, it has been used flexibly and carefully in particular treatment regimens rather than in a blanket, uniform fashion for everyone. AZT may be appropriate only for patients in whom viral replication is the main problem. "That's not East Coast opinion or West Coast opinion, but a simple fact of what is in the scientific literature." As for its use with asymptomatic people, West Coast logic is that AZT is most toxic with the sickest patients, so it will be least toxic with asymptomatic people: "A lot of the problem with AZT is that we're using it at the wrong time with the wrong people" (in Douglas 1989, 26). Delaney recommends early use: "The drug is *not* a poison. We do ourselves a disservice by starting from that premise or trying to prove it's a poison." Speaking for the AIDS community and arguing that people should at least be given options, Delaney values personal autonomy and individual experience.

Familiar with the script of AIDS debates, Delaney anticipated "East Coast" objections. Joseph Sonnabend, M.D., an academic researcher who now practices as a private AIDS physician in New York and has pioneered the Community Research Initiative there, at this point produced the argument that Delaney had been anticipating. Sonnabend contended that AZT *is* a "poison"; because it is not selective, it inhibits both "the replication of the virus *and* the replication of the host cell DNA. . . . It will effectively terminate chains of host DNA as well as viral DNA." In antiviral research, he observed, selectivity has traditionally been an important criterion; "For some reason in the case of HIV these principles have been abandoned." He repeated his continuing criticism that the AZT multicenter trials did not control for the quality and kind of medical care the patients received, and reasserted his thesis that medical care is "the most important determinant of life and death in the short term. . . . This includes *pneumocystis* prophylaxis, but that's not the only thing" (see Sonnabend 1989). He noted that there are still no federal guidelines for overall AIDS patient management.

In Metroka, Delaney, and Sonnabend we can identify a spectrum of views on AZT. Metroka, based on the rules of good scientific evidence, is ready to endorse AZT within the limits of the data. Sonnabend, often described as an independent thinker (not always a compliment in private medicine), bears the burden—no doubt at times a tiring one—of questioning the rules themselves—that is,

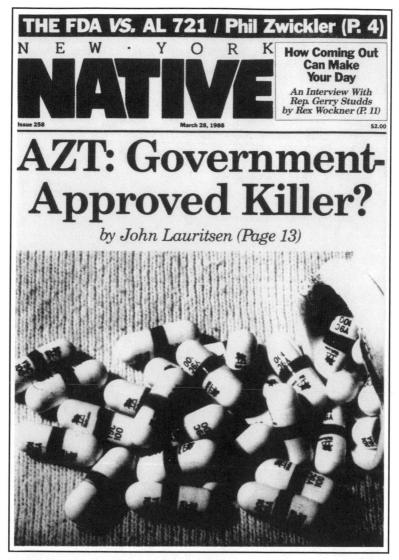

THE FDA *VS.* AL 721 / Phil Zwickler (P. 4)

N   E   W   ·   Y   O   R   K

NATIVE

**How Coming Out
Can Make
Your Day**

*An Interview With
Rep. Gerry Studds
by Rex Wockner (P. 11)*

Issue 258   March 28, 1988 $2.00

AZT: Government-
Approved Killer?

by John Lauritsen (Page 13)

Cover of the *New York Native*, March 28, 1989.

what constitutes good scientific evidence. To do so, he must repeatedly speak his piece and get his resistance on the record. Delaney, a
pragmatist with an agenda that calls for expanded options, predict-
ably calls it counterproductive to harp on the problems of the origi-

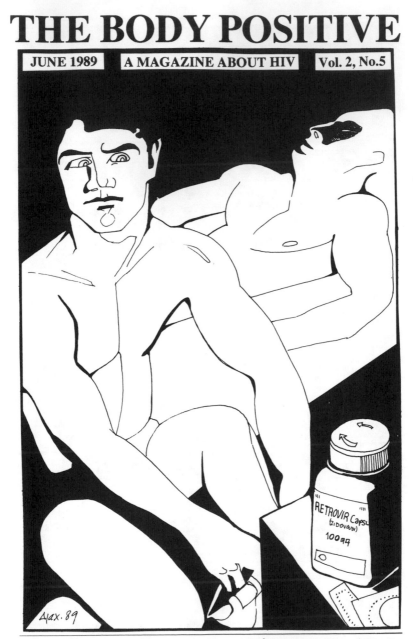

THE BODY POSITIVE

JUNE 1989 | **A MAGAZINE ABOUT HIV** | **Vol. 2, No.5**

DATING -SEX -TELLING   **versión en Español en página 4**

Cover of *The Body Positive*, June 1989. Published by Body Positive, New York City.

nal multicenter study—"a little like having study groups on the Council of Trent." Delaney, as a representative of the activist community, signals his own credentials not by invoking science but by asserting that AZT is now endorsed by "people who hate Burroughs-Wellcome as much as I do—and there's no one who hates them more than I do" (in Douglas 1989, 33).

Next to speak was Michael Lange, M.D., an associate professor at Columbia University College of Physicians and Surgeons, who contributed his own laundry list of AZT's problems. Like Sonnabend, Lange believes the claims for the drug are not supported by good evidence; he cites seven criticisms, all of which he has made at the FDA hearings ("Therapeutic Drugs for AIDS," 1988). Nor is Lange convinced that AZT has any antiviral effect or works in human beings at all. He notes that the truth about AZT is crucial to obtain, not only for the sake of patients in the United States, but because Third World countries are now discussing whether to invest in it at its unaffordable price. Not to be "strict with ourselves" here (in the First World) is to play "into the hands of the military-industrial complex at a tremendous cost" (in Douglas 1989, 30).

At this point Ronald Grossman, M.D., a private internist in New York, entered as peacemaker:

> *Pace, pace* Joe, *pace* Michael. Despite all the disclaimers to the contrary, I've heard war words, and I think we need to avoid that desperately in this situation. . . . I am a clinician, not a researcher. I see real live patients who are achieving real live benefits from this drug—and plenty of them who get toxic effects, just as we see with every other drug we use in medicine. (in Douglas 1989, 30)

Grossman then related the apparent benefits of AZT to the familiar tripod model in which the three determinants of health are (1) medical care, including the doctor-patient relationship, technology, and medications; (2) self-care, including diet, rest, and life-style; and (3) spirituality and positive thinking. "If we're offering hope with AZT," he concluded, "that strengthens the clinical benefit."

Michael Callen is an AIDS activist who was diagnosed with AIDS in 1982; he is a founder of the PWA Coalition, a writer and musician, and a patient of Joseph Sonnabend. He now emphasizes the difficulty of expressing skepticism in the context of AIDS theory. Despite his own attacks on AZT, Callen typically manages to construct a self-reflexive commentary that is always in some respects about the politics and purposes of speaking: "I realize," he said at the conference, alluding to his experiences speaking around the country, "that you can't breeze into a city or a group of people whose buzzers are going off all the time and say, cavalierly, as I have said, AZT is Drano in pill form, that it is poison." He said that he's often told to cool out, but he believes the "rational" procedure is to look at both sides and decide. ("In prison, I've been told, if somebody with AIDS chooses not to take AZT, they're diagnosed as suffering from HIV dementia!") "In response to Dr. Grossman's curious point that AZT is hope," he added, "let's give people some non-toxic substances that are also offering hope" (in Douglas 1989, 31).[9]

Neither Delaney nor Callen is a trained medical professional; nevertheless, both have extensive knowledge and conceptual grasp of drug actions and are influential in their communities. On AZT, they are on opposite sides, and at this point in the debate Delaney lost patience with Callen's skepticism and self-reflexivity: "This isn't an argument about how many angels can dance on the head of a pin. People's lives hang in the balance of this decision." The discussion then became heated. Lange demanded to be shown one good study on AZT. Delaney and Grossman responded that many good papers were presented at the Stockholm conference. Lange asserted that those papers also showed that AZT does nothing for the wasting syndrome that characterizes AIDS in Africa and is therefore probably *not* an antiviral drug. Delaney retorted that *of course* AZT is an antiviral: "I can't find many people outside this room or outside of this table who suggest that that's the case" (in Douglas 1989, 33). Laura Pinsky, the moderator, intervened at this point by asking Metroka, as coeditor of a treatment newsletter, to sum up AZT's efficacy. "Reviewing the literature," Metroka responded, "I would say that AZT has efficacy, while the drug is clearly not for everyone." Pinsky then commented that the conference organizers deliberately set up this panel to reflect differing opinions, but the audience should know that Sonnabend, Lange, and Callen are very much in the minority. Someone from the audience shouted, "That doesn't mean they're wrong!"

Donald Kotler, M.D., associate professor of clinical medicine at St. Luke's-Roosevelt, suggested that AZT can be viewed as a "negative cofactor" in disease development, just as the existence of other viruses or infections is a "positive co-factor":

I think we've all fallen into the trap of believing that a prospective randomized placebo-controlled double-blind trial is the ultimate arbiter of truth. In point of fact, it's not. . . . I would think that as a physician there is perhaps a better truth, and the better truth is one's own experience. My experience is that in some people AZT really has worked very well, and in some people it has not. [He provides two examples.] I feel looking at both those experiences that the personal experience to me is irrefutable. The FDA does not see it, but the FDA doesn't see my patients, they look at report forms. (in Douglas 1989, 36)

Kotler's position here is almost classically distinct from Metroka's; whereas Metroka relies on published controlled aggregate data to certify—and guard against—less formal perceptions and reports of success, Kotler trusts in the empirical lessons of observed clinical experience. The bottom line of this "better truth" is not necessarily that AZT is always good, but that only the physician and patient can determine whether, for individual patients, AZT "really has worked." Direct clinical observation and individual experience are taken as unique sources of knowledge.

These comments more or less represent the universe of discussion regarding zidovudine at the 1988 Columbia conference.[10] Developments since the conference, based on data from such studies as that of Volberding et al. (1990), described above, have included the widespread early use of low-dose AZT. Specific questions remain, above all the issues of when, precisely, intervention should occur— what Friedland (1990) calls the "golden moment." One might argue that further data are the only thing needed, but, as I have tried to suggest, the AZT argument is not entirely about data.

One central subtext is power. Callen and Delaney demonstrate a significant feature of AIDS activism: the refusal of patients to be patients, and the corollary determination of research subjects to be speaking subjects. It is said of Western medicine that the patient comes to the physician's office with an illness but leaves with a disease. Disease is thus taken to represent the medical model, and illness the patient's subjective experience; the primary-care physician plays a crucial role in mediating between individual subjective experience (illness) and the objective system of biomedical science (disease). But here we also see an insistence that patients' interests must, in some contexts, be treated as distinct from the interests of physicians. This insistence recalls anthropologist Michael T. Taussig's

(1980) argument that the clinician's attempt to understand the patient's cultural construction of illness—the "native's point of view"—does not adequately recognize the institutional power structures that traverse the clinical experience. Despite the physician's desire to identify with the patient, "there will be irreconcilable conflicts of interest and these will be 'negotiated' by those who hold the upper hand, albeit in terms of a language and a practice which denies such manipulation and the existence of unequal control." The issue, Taussig argues, is not "the cultural construction of clinical reality" but the "clinical construction of culture."

A second underlying narrative concerns technology and equity. Lange's invocation of responsibilities toward the Third World is one place where this surfaces explicitly. But, as Paul Douglas and Laura Pinsky (1989) argue, concerns about equity have implicitly shaped many AIDS debates. The resistance to early intervention in New York, they suggest, in part reflects the sorry state of health care delivery and social support services. It is not responsible, some activists and clinicians believe, to advocate early intervention—whether early use of AZT, PCP prophylaxis, or simply good nutrition—when such treatment is unobtainable by most HIV-infected people. While Friedland (1990) emphasizes the obvious medical benefits of early-use AZT, he also acknowledges the staggering economic and policy implications of adopting early intervention strategies (see Arno et al. 1989).

Appeals to established scientific fact function to support positions already held. At one extreme are officials and scientists upholding the value of strict clinical trial protocols; at the other, AIDS activists who likewise support their views by referring (as Delaney does) to the "simple fact of what is in the scientific literature." As Karl Mannheim long ago argued in *Ideology and Utopia,* out of prolonged social debates arise steadfast defenders and passionate challengers of the status quo who are equally skilled at constructing different interpretations of the "facts" about the world to support their cause. To dispute these "facts" becomes increasingly difficult over time, because the gradual acceptance of one interpretation tends also to naturalize the processes and assumptions through which it was arrived at. At the same time, as observers of scientific practice have argued, once "facts" are widely accepted, they become synonymous with reality and truth and in some sense render the quest for truth irrelevant or uninteresting (Fleck 1979; Knorr-Cetina 1981; Latour and Woolgar 1986).

Latour and Woolgar go further, however, when they suggest that the authority to define reality is reinforced by an intersection of interests; "reality," indeed, may be defined as that set of statements that has become too costly to give up. This takes us to the heart of the impatience that "practical" people often express toward "theory"— Delaney's charge, for instance, that to continue to criticize the early AZT data is "like holding study groups on the Council of Trent," or his assertion that "this isn't an argument about how many angels can dance on the head of a pin." This is always a dismissive comment, designed to characterize the opponent's argument as scholastic when "people's lives hang in the balance of this decision." But it is also an enactment of power that asserts that "reality"—the set of statements too costly to give up—is now taken as settled and is no longer vulnerable to questions of abstract theory.

Conclusion

> *I'm wanting something new*
> *Say, have you got a clue*
> *where I can get ahold of some*
> *Compound Q?*
>
> *Compound Q—Antabuse—Acyclovir:*
> *I got the drug—right—here!*
> —Ron Goldberg, "Fugue for Drug Trials"

In October 1988, when activists stormed the FDA, two people with AIDS in the United States were dying every hour. By May 1990, when activists targeted the NIH, the ACT UP poster claimed "ONE AIDS DEATH EVERY 12 MINUTES." As the second decade of the epidemic begins, and the shift among activists from prevention to treatment grows more intense and fraught with controversy and responsibility, how to have theory in an epidemic becomes an even more crucial question. AZT—like Compound Q and other drugs—held out, for a brief time, the promise of a magical cure for HIV—a cure that AIDS theory had long since declared impossible and that was at odds with the provisional, partial vision of science and truth that I have been attributing to AIDS activism. This makes the horse-racing metaphor

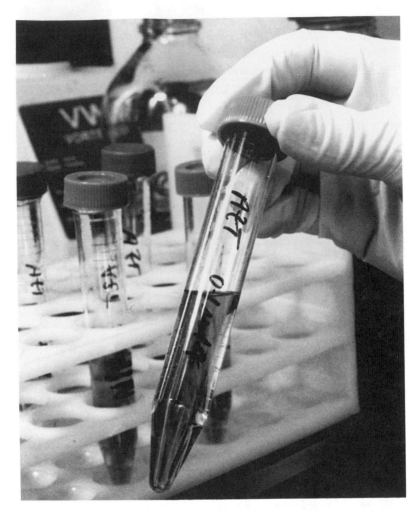

AZT is used to hold out hope for AIDS vaccines and cures, yet simultaneously to assert the immense difficulty, even impossibility, of developing them. An article in *Newsweek*, for example, calls the virus a "moving target" (Sharon Begley, with Mary Hager and Mark Starr, "Moving Target: Searching for a Vaccine and a Cure," November 24, 1986, p. 36). Photograph © Ira Wyman; used with permission.

underlying "Fugue for Drug Trials" all the more appropriate, however, for, as both the history of science and the *Racing Form* testify, magical things sometimes happen, and no movement can or should fully arm itself against hope or fully repress the desire that the unful-

Still from video *Seize Control of the FDA* (Bordowitz and Carlomusto, 1988).

fillable will be fulfilled. AIDS treatment activism guarantees that long-shots as well as favorites, proletarians as well as bluebloods, will all have their chance to run.

Though resistance to AZT has faded, other new and controversial drugs will appear to take its place. In one form or another, these divisive debates and seemingly irreconcilable positions about research and treatment are here to stay. Although what seem to be mutually exclusive discourses and world visions have many points of contact, coalition, and negotiation, their differences cannot be transcended by commonsense assertions about what is true, natural, or human, or by the eventual emergence of apparently consensual truths about HIV, AZT, or other specific controversies of the epidemic. These apparent resolutions in such a crisis can be neither stable nor permanent.

Yet, as Meurig Horton (1989, 171) argues, innovative structures like community-based AIDS treatment research programs offer "a social and theoretical space where the possibilities of research and treatment can be thought differently." Such programs, like the debates that generated them, furnish lessons about the rules, conventions, and values that anchor the production of knowledge and determine how "truth" in any given context will be decided. This will

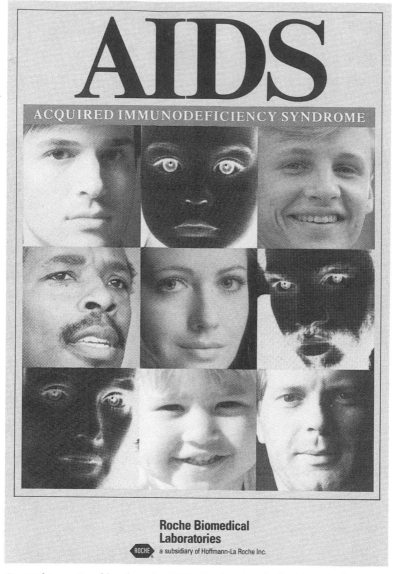

Cover of promotional brochure for AIDS-related testing. Roche Biomedical Laboratories, a subsidiary of Hoffman-La Roche, Inc., 1988.

be useful as the struggle broadens, as it eventually must, to challenge the health care system itself.

The debate about zidovudine, as sketched here, makes clear that many parties have information and analyses that are crucial to understanding and making choices about treatment. Although physicians and scientists have unique and valuable contributions to make, they are not inherently better informed than AIDS activists, nor is their knowledge more complicated than that which informed patients can bring to the treatment scene. It is also clear that contradictory evidence and widely divergent interpretations exist, though these are not always neatly identifiable among categories of professional training. As I have argued elsewhere, we draw upon diverse cultural resources to make sense of a complex and devastating crisis (Treichler 1988). The AIDS epidemic and the clinical reality of HIV infection and AIDS are intrinsically complicated and can be simplified only for specific strategic purposes. What is incontrovertible is that the volatile interactions entailed by these broadly inclusive debates — in both the short term and the long term — will benefit not only people with HIV infection but the culture as a whole. For they involve significant renegotiations of the geography of cultural struggle — of sources of biomedical expertise, relationships between doctor and patient, relationships of the general citizenry to science and to government bureaucracies, and debate about the role and ownership of the body. Not only are the basic definitions and self-images of these constituencies at stake but also the institutional and cultural structures that shape their relations to each other and their relative empowerment and effectivity within the culture as a whole.

The struggle for the right to preserve health, like that for the right to experience pleasure, is founded on a political and theoretical analysis of the body — how it works, what it experiences, and how it exists in society. Community-based treatment research does not make treatment decisions easier, but it enables them to take place in a context radically different from what was available a decade ago. And it provides a way of engaging, as a lived reality, the question of how many angels can dance on the head of a pin. This is what theory in an epidemic requires; it is one way to begin to put one's body where one's head is (or vice versa), to show courage when too many people have already died, and thus try to come to terms with what Donna Haraway (1989, 32) calls "the problematic multiplicities of postmodern selves." Organized resistance to activist proposals will require continuing struggle, and ultimately a direct engagement with the ineq-

Safer sex according to Jean Genet. Still from *The Pink Pimpernel* (Greyson 1989).

uities of the U.S. health care system. In the meantime, the struggles over AIDS treatment and the cultural forces that shape them can be seen as a narrative about the evolution of a radically democratic technoculture and about whose rules, in this democracy, we are to live by.

NOTES

For information about AIDS treatment activism and ongoing discussions about the AIDS epidemic, I am indebted to Gregg Bordowitz, Jean Carlomusto, Douglas Crimp, John Erni, Jamie Feldman, Jan Zita Grover, Bob Huff, Cary Nelson, Laura Pinsky, Andrew Ross, Joseph Sonnabend, and Simon Watney.

1.   Sung to the tune of "Fugue for Tinhorns" from the musical *Guys and Dolls,* music and lyrics by Frank Loesser; new lyrics by Ron Goldberg, copyright 1989. Performed at the ACT UP talent show, New York, March 30, 1989. Used with permission.

2.   See Bohne, Cunningham, Engbretson, Fortunate, and Harrington, *Treatment and Data Handbook* (1989, 47, 57). Informed consent provisions in different countries create widely varying standards for giving patients information and for keeping confidentiality. The United States has more stringent requirements for informing patients of risks and benefits of drugs and medical procedures than any other country; at the same time, other industrialized countries leave medical care and drug development less vulnerable to free enterprise. Horton (1989) describes the clash that occurred when medical researchers organizing clinical trials of AZT in England refused to share detailed information with their well-informed potential clientele. Other problems are created by record-keeping practices, which jeopardize confidentiality about

HIV status. The French do not tend to provide much information to patients, but they also do not require much from them, and would typically not chart potentially damaging test results.

3. In September 1990, the U.S. Congress mandated that all NIH research on human subjects include women as well as men.

4. These coalitions are not, of course, unproblematic (Kolata 1988b). The unvarnished profit motive comes as a shock to many liberals and activists: a faster approval process "translates into lower research costs and quicker profits" for pharmaceutical companies, a Biochem founder told Toughill (1989): "That's why Biochem chose to research AIDS drugs." Similarly, according to the cover story in *Business* by Franklin et al. (1987), the problem with vaccines is their poor profit margin—they are needed only once. Yet because Wall Street continues to debate whether the epidemic has "topped out" and how to separate "hype from hope" about AIDS drugs (Mahar 1989), its analysts are motivated to look closely and skeptically at scientific data, media coverage, and other forms of conventional wisdom (see "AIDS and 1962" 1988; Franklin et al. 1987; Ricklefs 1988; Maher 1989). While AIDS activists are willing to use the profit motive to their advantage, they do not trust it (any more than they trust the media with which they have learned to work so effectively). *Rockville Is Burning* (Huff and Wave 3 1989) satirizes Wall Street's pleasure with the "cocktail approach" to AIDS treatments in which a whole range of products will be tried. "This is a recession-proof growth industry," one robber baron tells another over drinks. "This thing could be bigger than skin care." A poster by the art collective Gran Fury, meanwhile, quotes a representative of the pharmaceutical company Hoffman LaRoche explaining why the company won't be developing AIDS drugs: "It's big—but it's not asthma" (*Art Forum,* October 1989).

5. Self-care is a long American tradition ("Each man his own doctor" ran a nineteenth-century aphorism) that has gained new ground since the civil rights, women's liberation, and consumer movements of the 1960s and 1970s. Peer support groups have also exploded during this period, some modeled on the twelve-step program of Alcoholics Anonymous, others adopting different formats and goals. Self-help and self-treatment have been perhaps most fully developed within the women's health movement, and it is to books like *Our Bodies, Ourselves* (Boston Women's Health Book Collective, 1984) that some AIDS treatment activists acknowledge their greatest debt. At the same time, many women AIDS activists may identify with the movement more as gay people than as feminists, whom they perceive as less than helpful in the AIDS crisis (see ACT UP/NY Women and AIDS Book Group 1990). Information about and useful distinctions among self-help, validated, nonvalidated, and unorthodox therapies are given by Cassileth and Brown (1988) and Freedman and McGill Boston Research Group (1989). Josh Gamson (1989) attempts to identify the features that make AIDS activism distinct as a new social movement.

6. Kramer is currently a member of ACT UP. His 1987 speech reflects his longstanding view that the federal effort against AIDS is unspeakably inadequate. As early as 1982, for example, he wrote that "studies are constantly announced and undertaken by people who have only the vaguest notions of how we live." This quote serves as an epigraph in ACT UP's (1989) *National AIDS Treatment Research Agenda.* Tensions between official and activist perspectives on the AIDS drug approval process are addressed by Bishop (1987), Boffey (1988b, 1988c, 1988d), Chase (1988), Douglas and Pinsky (1989), Geitner (1988), Goldstein and Massa (1989), Gross (1987), James

(1989a, 1989b), Kingston (1990), Leary (1988), and Rothman (1987). Erni (forthcoming) explores tensions in AIDS discourse around the topic of "a cure."

7. The agenda lists twelve principles for a new AIDS drug testing system; proposes new models and suggestions for speeding Phase I safety trials, pilot efficacy trials, new treatment protocols, and postmarketing surveillance; sets priorities for clinical research into several dozen specific AIDS drugs; and highlights "Five Drugs We Need Now" and "Seven Treatments We Want Tested Faster." The final section on AIDS drug disasters lists nine drugs for which development has been delayed, mismanaged, neglected, or prevented. While this document is directed toward the FDA system, complementary publications address decision making for individuals; see, for example, *Deciding to Enter an AIDS/HIV Drug Trial* (1989) and the excellent educational video *Work Your Body*, produced for the Gay Men's Health Crisis Living With AIDS series (Bordowitz and Carlomusto, 1988). For a more detailed analysis of the problems for women and drug trials, see ACT UP/NY Women and AIDS Book Group (1990). For a discussion of why some people resist treatment and some organizations do not emphasize alternatives, see Douglas and Pinsky (1989). ACT UP also has an Alternative and Holistic Treatment Subcommittee, but this is not my focus here.

8. In this epigraph, "the gift of Screws" signifies the mechanical press that is used to extract fragrant oils from flower petals. In the summer of 1989 I was telling a young man in Illinois the saga of trichosanthin, an AIDS drug popularly known as Compound Q, a highly purified form of a protein derived from the root of *Trichosanthes kirilowii*, a Chinese cucumber (see McGrath et al. 1989). I said that in vitro the drug seemed to be relatively selective, killing HIV-infected cells but not uninfected cells; still being tested in both approved and underground trials, Compound Q had nonetheless been hailed in the media as the latest cure for AIDS. He was delighted with this story. "Wouldn't you know the cure for AIDS would be a cucumber — something natural," he said, "and not some horrible toxic chemical." Born a cucumber perhaps, trichosanthin the drug is "the gift of Screws" — of laboratory operations and biochemical manipulations and capital investment and human effort. Moreover, initial tests in vivo reveal it to have highly potent, complex effects on HIV-infected people, with numerous side effects. Its further development will depend on technology, and even then it will not likely be a "cure." Dickinson's poem, then, can be read as a statement about the natural as almost always already technological. On Compound Q, see Goldstein and Massa (1989), Kingston (1990), and *Treatment Issues* (3 [October 30, 1989], 9).

9. An interesting set of discourses converges around this question of whether AZT is a "poison." A technical discourse may define a poison simply in terms of its effects on cells and DNA replication. That many antiviral drugs have broad effects on cellular replication is one of their generic difficulties. But the charge of "poison" is also part of a readily available cultural narrative about the poisoning of the body, the environment, the earth; about the poisoning of the natural by the technological. The British treatment journal *Positively Healthy* (see Marshall 1989) makes a sustained critique of AZT and other potent medications on the grounds that they poison the body with the same kind of systemic toxins that poison the earth. What is needed is to rebuild the body through intensive work with nutritive substances, not further destroy its fragile structures. Nevertheless, the journal provides detailed and timely treatment information, and from this highly engaged position contests the terrain of AIDS discourse in technical analytic terms.

This is quite different from, say, some New Age articulations of "the natural." A collection of "channeled teachings" on AIDS encourages the use of natural healing processes, inner guidance, and internal "chemical inducers" to boost the immune system (Spirit Speaks 1987), but its discourse is generic and out of touch with the details of AIDS research and treatment. Despite repeated charges of traditional medicine's misconceptions, in the end, the book represents AIDS as something that happens on the "Earth plane"—indeed, *needs* to happen there, whatever that means—and is of no concern: "Those within the Spirit dimension will not interfere in *any way* with the Earth plane" (p. 168). At the same time, though condoms, drugs, and vaccines are represented as ultimately illusory and shallow, they are nonetheless recommended by the spirits, who also counsel individuals not to hassle their doctors with the truth, which the doctors won't understand. One of the lessons of AIDS activism is that a challenge to any given version of "truth" will be most effective when it is anchored in a coherent counternarrative. Yet in the case of channeling, an entire alternative referential apparatus is constructed to no earthly avail. Its theories and therapies refuse an orthodox biomedical worldview in favor of an alternate universe that has no observable consequences and places virtually no burden on institutions such as the FDA, existing as they do only on the "Earth plane." Contrast this with the oppositional and contentious universe of AIDS treatment activism, which, whatever its internal disagreements, regularly produces an ambitious agenda for established institutions.

10. Subsequently, the *New York Times* reported on August 18, 1989, that findings of a thirty-two-center study indicated that AZT will help "AIDS cases with virus but no symptoms," making it half as likely for those receiving the drug to develop symptoms (Hilts 1989, 1). The secretary of Health and Human Services, Dr. Louis Sullivan, announced the findings in Washington: "Today we are witnessing a turning point in the battle to change AIDS from a fatal disease to a treatable one." Fauci was quoted in the article as saying that the findings made it important for people to get themselves tested. Dr. Samuel Broder, confirming what claimed to be the study's definitive nature, stated in the article that the issue regarding asymptomatics "has now been resolved" (p. 12). But other scientists and AIDS groups emphasized that the *Times* story was little more than a reprint of Burroughs-Wellcome's press release and urged caution until published data were available. And on October 16, 1989, this full-page ad appeared on page 20 of the *New York Times*:

> BEFORE YOU
> TAKE AZT AGAIN,
> READ THE
> NOVEMBER
> ISSUE OF
> SPIN

In that issue of *Spin*, Celia Farber's regular "AIDS: Words from the Front" column, titled "Sins of Omission: The AZT Scandal," reviewed loopholes in the approval process for AZT and quoted a number of dissident scientists. Dr. Peter Duesberg, for example, a retrovirologist at the University of California at Berkeley, told Farber (1989, 117) that asymptomatic seropositive people who take AZT "are running into the gas chamber."

In another postconference judgment, John Lauritsen (1989, 17), in his continuing series on the *New York Native* on AZT as "poison by prescription," negatively reviewed the Columbia conference: "On the whole it was a flop." He described Metroka's talk as "almost inhuman in its glibness" and called Delaney's talk "a hard-sell pitch for AZT."

Lauritsen, in a sense, has a role in the script too—that of the revealer of the conspiracy. He describes being harassed at the conference; denials are taken as confirming evidence. The compelling nature of the *Native*'s theories is evident in its fierce defenders. Yet one of my friends was given a formal prescription by his therapist not to read it; the therapist promised that if any major developments occurred, he would communicate them. Many others would doubtless credit the *Native* with keeping them alive through this terrifying period. This is yet one more example of the hazardous path among sources that people must learn to weave for themselves.

REFERENCES

ACT UP. *A National AIDS Treatment Research Agenda* (New York: ACT UP, June 1989).

ACT UP. *A Critique of the AIDS Clinical Trials Group* (New York: ACT UP, May 1990).

ACT UP/NY Women and AIDS Book Group. *Women, AIDS and Activism* (Boston: South End Press, 1990).

"AIDS and 1962." Editorial, *Wall Street Journal* (July 14, 1988a), 26.

"An AIDS Crisis Proposal." Editorial, *Wall Street Journal* (June 15, 1988b), 24.

Altman, Lawrence K. "Inhaled Drug Is Found to Benefit Against Pneumonia in AIDS Cases." *New York Times* (June 15, 1988), A21.

Arno, Philip S., et al. "Economic and Policy Implications of Early Intervention in HIV Disease." *Journal of the American Medical Association,* 262 (1989), 1493–98.

Bayer, Ronald. *Homosexuality and American Psychiatry: The Politics of Diagnosis* (New York: Basic Books, 1981).

Bishop, Katherine. "Frustrated AIDS Patients Devise Their Own Therapies." *New York Times* (March 17, 1987), 16.

Boffey, Philip M. "AIDS Panel Wants Wider Drug Tests." *New York Times* (February 21, 1988a), 32.

Boffey, Philip M. "At Fulcrum of Conflict, Regulator of AIDS Drugs." *New York Times* (August 19, 1988b), 12.

Boffey, Philip M. "FDA Will Allow AIDS Patients to Import Unapproved Medicines." *New York Times* (July 25, 1988c), 1, 10.

Boffey, Philip M. "Low AIDS Budget of FDA Said to Slow Drug Approval." *New York Times* (February 20, 1988d), 7.

Bohne, John, Tom Cunningham, Jon Engebretson, Ken Fortunato, and Mark Harrington. *Treatment and Data Handbook: Treatment Decisions* (New York: ACT UP, 1989).

Bordowitz, Gregg, and Jean Carlomusto, producers. *Work Your Body* (video, Living With AIDS series) (New York: Gay Men's Health Crisis, 1988).

Bordowitz, Gregg, and Jean Carlomusto, producers. *Seize Control of the FDA* (video, Living With AIDS series) (New York: Gay Men's Health Crisis, 1989).

Boston Women's Health Book Collective. *The New Our Bodies, Ourselves* (New York: Simon & Schuster, 1984).

Callen, Michael, ed. *Surviving and Thriving with AIDS,* 2 vols. (New York: PWA Coalition, 1987).

Callen, Michael. "AIDS and Passive Genocide: 30,534 Unnecessary Deaths from PCP Due to a Scandalous Failure to Prophylax." Testimony given at FDA hearing concerning the approval of aerosol pentamidine as prophylaxis against PCP, May 1, 1989. *AIDS Forum,* 2 (May 1989), 13–16.

Callen, Michael, and Richard Berkowitz (with assistance from Dr. Joseph Sonnabend and Richard Dworkin). *How to Have Sex in an Epidemic* (New York: News from the Front Publications, 1983).

Cassileth, Barrie R., and Helene Brown. "Unorthodox Cancer Medicine." *CA-A Cancer Journal for Clinicians*, 38, 3 (1988), 176–86.

Centers for Disease Control. "CDC Guidelines for Prophylaxis against PCP for Persons Infected with HIV." *Morbidity and Mortality Weekly Report*, 38, S-5 (1989), 1–9.

Chase, Marilyn. "U.S.-Sponsored AIDS Drug Trials to Include Private Doctors' Efforts." *Wall Street Journal* (November 23, 1988), B–3.

Crimp, Douglas. "How to Have Promiscuity in an Epidemic." In *AIDS: Cultural Analysis/Cultural Activism*, ed. Douglas Crimp (Cambridge: MIT Press, 1988), 237–71.

Crimp, Douglas, with Adam Rolston. *AIDS Demo Graphics* (Seattle: Bay, 1990).

Daniel, Herbert. *Life Before Death* (Rio de Janeiro: Tipografia Joboti, 1989).

Deciding to Enter an AIDS/HIV Drug Trial (New York: AIDS Treatment Registry, Inc., Summer 1989).

Douglas, Paul, Harding, ed. *AIDS: Improving the Odds 1988.* Transcript of proceedings, Columbia Gay Health Advocacy Project annual conference, Columbia University, November 19, 1988 (New York: Columbia Gay Health Advocacy Project, March 1989).

Douglas, Paul, and Laura Pinsky. "AIDS and Needless Deaths: How Early Treatment Is Ignored." Presented at the Seminar on Sex, Gender, and Consumer Culture, New York Institute for the Humanities (April 14, 1989).

Erni, John. " 'Curing AIDS': Biomedical Discourses, the Media, and the Politics of the 'Diseased Body.' " *Communication* (forthcoming).

Farber, Celia. "Sins of Omission: The AZT Scandal." *Spin*, 5 (November 1989), 40.

FitzGerald, Frances. *Cities on a Hill: A Journey Through Contemporary American Cultures* (New York: Simon & Schuster, 1986).

Fischl, M. A., et al. "The Efficacy of Azidothymidine (AZT) in the Treatment of Patients with AIDS and AIDS-Related Complex." *New England Journal of Medicine*, 317, 4 (1987), 185–97.

Fleck, Ludvik. *Genesis and Development of a Scientific Fact*, ed. Thaddeus J. Trenn and Robert K. Merton (Chicago: University of Chicago Press, 1979). (First published 1936.)

Franklin, Patricia, et al. "The AIDS Business." *Business*, 2 (April 1987), 42–47.

Freedman, Benjamin, and McGill Boston Research Group. "Nonvalidated Therapies and HIV Disease." *Hastings Center Report* (May/June 1989), 14–20.

Friedland, Gerald H. "Early Treatment for HIV: The Time Has Come." *New England Journal of Medicine*, 322 (April 5, 1990), 1000–1002.

Gamson, Josh. "Silence, Death, and the Invisible Enemy: AIDS Activism and Social Movement 'Newness.' " *Social Problems*, 36 (October 1989), 351–67.

Geitner, Paul. "Desperation Draws Victims to Try Unapproved Drugs." *Champaign Urbana News-Gazette* (June 19, 1988), B–4.

Gieringer, Dale. "Twice Wrong on AIDS." *New York Times* (January 12, 1987), A21.

Goldstein, Richard, and Robert Massa. "Compound Q: Hope and Hype; the Making of a New AIDS Drug." *Village Voice* (May 30, 1989), 29–34.

Goodfield, June. *Quest for the Killers* (New York: Hill & Wang, 1985), 51–97.

Greyson, John. *The Pink Pimpernel* (video) (Toronto: John Greyson, 1989). (Distributed by V/Tape, Toronto.)

Gross, Jane. "AIDS Victims Grasp at Home Remedies and Rumors of Cures." *New York Times* (May 15, 1987), 13.

Haire, Doris. *How the F.D.A. Determines the "Safety" of Drugs—Just How Safe Is "Safe"?* (report released to the Congress of the United States) (Washington, D.C.: National Women's Health Network, 1984).

Haraway, Donna. "The Biopolitics of Postmodern Bodies: Determinations of Self in Immune System Discourse." *Differences,* 1, 1 (1989), 3–43.

Harrington, Mark. "What I Said at the FDA Advisory Committee Meeting on Parallel Track." *ACT UP Reports,* 1 (September/October 1989), 5–6.

Harrington, Mark. *A Critique of the AIDS Clinical Trials Group,* ed. Ken Fortunato (New York: ACT UP Treatment and Data Committee, May 1, 1990).

Hilts, Philip J. "Drug Said to Help AIDS Cases with Virus but No Symptoms." *New York Times* (August 18, 1989a), 1.

Hilts, Philip J. "Wave of Protests Developing on Profits from AIDS Drug." *New York Times* (September 16, 1989b), 1.

Horton, Meurig. "Bugs, Drugs and Placebos: The Opulence of Truth, or How to Make a Treatment Decision in an Epidemic." In *Taking Liberties: AIDS and Cultural Politics,* ed. Erica Carter and Simon Watney (London: Serpent's Tail, 1989), 161–81.

Huff, Bob, and Wave 3. *Rockville Is Burning* (video) (New York: Bob Huff and Wave 3, 1989).

James, John S. "The Drug-Trials Debacle, Part 1: What to Do About It." *AIDS Treatment News,* 77 (April 21, 1989a), 3–6.

James, John S. "The Drug-Trials Debacle, Part 2: What to Do Now." *AIDS Treatment News,* 78 (May 5, 1989b), 4–8.

Jones, James H. *Bad Blood: The Tuskegee Syphilis Experiment—A Tragedy of Race and Medicine* (New York: Free Press, 1981).

Kingston, Tim. "Parallel Track." *San Francisco Bay Times* (May 1990), 4–5, 15.

Knorr-Cetina, Karin D. *The Manufacture of Knowledge: An Essay on the Constructivist and Contextual Nature of Science* (Oxford: Pergamon, 1981).

Kolata, Gina. "AIDS Patients and Their Above-Ground Underground." *New York Times* (July 10, 1988a), E32.

Kolata, Gina. "Odd Alliance Would Speed New Drugs." *New York Times* (November 26, 1988b), 9.

Kramer, Larry. "Taking Responsibility for Our Lives: Does the Gay Community Have a Death Wish?" *New York Native* (June 29, 1987), 37–40, 66–67.

Krim, Mathilde. "A Chance at Life for AIDS Sufferers." *New York Times* (August 8, 1986), A27.

Krim, Mathilde. "Making Experimental Drugs Available for AIDS Treatment." *AIDS Public Policy Journal,* 2 (1987), 1–5.

Latour, Bruno, and Steve Woolgar. *Laboratory Life: The Construction of Scientific Facts* (Princeton, N.J.: Princeton University Press, 1986).

Lauritsen, John. "The AZT Front." *New York Native,* 298 (January 2, 1989), 16–18. (Reprinted in *Poison by Prescription: The AZT Story,* a collection of 1987–89 articles from the *New York Native.*)

Leary, Warren E. "F.D.A. Pressed to Approve More AIDS Drugs." *New York Times* (October 11, 1988), C–5.

Levine, Robert J., and Karen Lebacqz. "Ethical Considerations in Clinical Trials." *Clinical Pharmacology and Therapeutics,* 25 (May 1979), 728–41.

Mahar, Maggie. "Pitiless Scourge: Separating Out Hype from Hope on AIDS." *Barron's* (March 13, 1989), 6–7, 16, 18, 20, 22–24, 26.

Mannheim, Karl. *Ideology and Utopia* (New York: Harcourt Brace Jovanovich, 1985). (First published 1936.)

Marshall, Stuart. "Don't Blame Me." *Positively Healthy,* 2 (March 1989), 13–14.

Massa, Robert. "Why AIDS Activists Target the FDA." *Village Voice* (October 18, 1987), 25.

McGrath, Michael S., et al. "GLQ223: An Inhibitor of Human Immunodeficiency Virus Replication in Acutely and Chronically Infected Cells of Lymphocyte and Mononuclear Phagocyte Lineage." *Proceedings of the National Academy of Sciences,* 86 (April 15, 1989), 2844–48.

President's Commission. *Final Report of the President's Commission for the Study of Ethical Problems in Medicine and Biomedical and Behavioral Research* (Washington, D.C.: U.S. Government Printing Office, 1983).

Presidential Commission. *Report of the Presidential Commission on the Human Immunodeficiency Virus Epidemic* (Washington, D.C.: U.S. Government Printing Office, 1988).

Ricklefs, Roger. "Gay-Rights Groups and Insurers Battle over Required AIDS Tests." *Wall Street Journal* (April 26, 1988), 41.

Rothman, David J. "Ethical and Social Issues in the Development of New Drugs and Vaccines." *Bulletin of the New York Academy of Medicine,* 63, 6 (1987), 557–68.

Russo, Vito. "State of Emergency: A Speech from the AIDS Movement." *Radical America,* 21, 6 (1988), 64–68.

Shilts, Randy. *And the Band Played On: People, Politics, and the AIDS Epidemic* (New York: St. Martin's, 1987).

Sonnabend, Joseph A. "Review of AZT Multicenter Trial Data Obtained under the Freedom of Information Act by Project Inform and ACT UP." *AIDS Forum,* 1 (January 1989), 9–15.

Spirit Speaks. *AIDS: From Fear to Hope: Channeled Teachings Offering Insight and Inspiration* (Miami: New Age, 1987).

Taussig, Michael T. "Reification and the Consciousness of the Patient." *Social Science and Medicine,* 14B (1980), 3–13.

"Therapeutic Drugs for AIDS: Development, Testing and Availability." Hearings before a subcommittee of the Committee on Government Operations, House of Representatives, 100th Congress, Second Session (April 28–29, 1988).

Torres, Gabriel. "New Therapies for PCP." *Treatment Issues,* 3 (December 5, 1989), 7–10.

Toughill, Kelly. "No Breaks on Price of New AIDS Drug, Firm's Founder Says." *Toronto Star* (April 3, 1989), A4.

Treichler, Paula A. "AIDS, Homophobia, and Biomedical Discourse: An Epidemic of Signification." *AIDS: Cultural Analysis/Cultural Activism,* ed. Douglas Crimp (Cambridge: MIT Press, 1988), 31–70.

"The Trials of AZT." Editorial, *Positively Healthy,* 2 (March 1989), 1.

Volberding, Paul A., et al. "Zidovudine in Asymptomatic Human Immunodeficiency Virus Infection: A Controlled Trial in Persons with Fewer than 500 CD4-Positive Cells per Cubic Millimeter." *New England Journal of Medicine,* 322 (April 5, 1990), 941–49.

Wechsler, Nancy. "Diary: FDA Action." *Radical America,* 21, 6 (1988), 71–72.

Young, Frank E., et al. "The FDA's New Procedures for the Use of Investigational New Drugs in Treatment." *Journal of the American Medical Association,* 239, 15 (1988), 2267–70.

Zonona, Victor F. "Bootstrap AIDS Research Giving Patients Active Role." *Los Angeles Times* (December 25, 1988), sec. 1, 1.

Selected AIDS Treatment Research Reports and Periodicals

ACT UP New York Reports. 496A Hudson Street, Suite G4, New York, NY 10014; (212) 989–1114.

AIDS Clinical Care. Massachusetts Medical Society, 1440 Main Street, Waltham, MA 02154; (617) 983-3800.

AIDS/HIV Experimental Treatment Directory. American Foundation for AIDS Research, 1515 Broadway, Suite 3601, New York, NY 10036; (212) 719-0033.

AIDS Targeted Information Newsletter: Abstracts and Critical Comments from Current AIDS Literature. Williams & Wilkins, 428 E. Preston Street, Baltimore, MD 21202.

AIDS Treatment News. John S. James, P.O. Box 411256, San Francisco, CA 94141; (415) 255-0588.

The Body Positive. Body Positive, 208 W. 13th Street, New York, NY 10011; (212) 633-1732.

Community Research Initiative. 31 W. 26th Street, New York, NY 10010; (212) 481-1050.

FDA Drug Bulletin. Circulation Dept. HFI-43, 3600 Fishers Lane, Rockville, MD 20857.

NIAID Clinical Trials Information Service. P.O. Box 6421, Rockville, MD 20850; toll-free hot line, (800) TRIALS-A.

PI Perspective. Project Inform; toll-free hot line, (800) 822-7422.

Positively Healthy: Transformation through Information. P.O. Box 71, Richmond, Surrey TW9 3DJ England.

PWA Coalition Newsline. PWA Coalition, Inc., 31 W. 26th Street, New York, NY 10011; (212) 532-0290.

Treatment Issues: The GMHC Newsletter of Experimental AIDS Therapies. Dept. of Medical Information, Gay Men's Health Crisis, 129 W. 20th Street, New York, NY 10011.

Hacking Away at the Counterculture
Andrew Ross

Ever since the viral attack engineered in November of 1988 by Cornell University hacker Robert Morris on the national network system Internet, which includes the Pentagon's ARPAnet data exchange network, the nation's high-tech ideologues and spin doctors have been locked in debate, trying to make ethical and economic sense of the event. The virus rapidly infected an estimated six thousand computers around the country, creating a scare that crowned an open season of viral hysteria in the media, in the course of which, according to the Computer Virus Industry Association in Santa Clara, California, the number of known viruses jumped from seven to thirty during 1988, and from three thousand infections in the first two months of that year to thirty thousand in the last two months. While it caused little in the way of data damage (some richly inflated initial estimates reckoned up to $100 million in downtime), the ramifications of the Internet virus have helped to generate a moral panic that has all but transformed everyday "computer culture."

Following the lead of the Defense Advance Research Projects Agency (DARPA) Computer Emergency Response Team at Carnegie-Mellon University, antivirus response centers were hastily put in place by government and defense agencies at the National Science Foundation, the Energy Department, NASA, and other sites. Plans were made to introduce a bill in Congress (the Computer Virus Eradication Act, to replace the 1986 Computer Fraud and Abuse Act, which pertained solely to government information) that would call for prison sentences of up to ten years for the "crime" of sophisti-

cated hacking, and numerous government agencies have been involved in a proprietary fight over the creation of a proposed Center
for Virus Control, modeled, of course, on Atlanta's Centers for Disease Control, notorious for its failure to respond adequately to the
AIDS crisis.

Media commentary on the virus scare has run not so much tongue-
in-cheek as hand-in-glove with the rhetoric of AIDS hysteria—the
common use of terms like *killer virus* and *epidemic;* the focus on
high-risk personal contact (virus infection, for the most part, is
spread on personal computers, not mainframes); the obsession with
defense, security, and immunity; and the climate of suspicion generated around communitarian acts of sharing. The underlying moral
imperative being this: You can't trust your best friend's software any
more than you can trust his or her bodily fluids—safe software or no
software at all! Or, as Dennis Miller put it on *Saturday Night Live,* "Remember, when you connect with another computer, you're connecting to every computer that computer has ever connected to." This
playful conceit struck a chord in the popular consciousness, even as
it was perpetuated in such sober quarters as the Association for Computing Machinery, the president of which, in a controversial editorial
titled "A Hygiene Lesson," drew comparisons not only with sexually
transmitted diseases, but also with a cholera epidemic, and urged attention to "personal systems hygiene." [1] Some computer scientists
who studied the symptomatic path of Morris's virus across Internet
have pointed to its uneven effects upon different computer types and
operating systems, and concluded that "there is a direct analogy with
biological genetic diversity to be made." [2] The epidemiology of biological virus, and especially AIDS, research is being studied closely to
help implement computer security plans, and, in these circles, the
new witty discourse is laced with references to antigens, white blood
cells, vaccinations, metabolic free radicals, and the like.

The form and content of more lurid articles like *Time*'s infamous
(September 1988) story, "Invasion of the Data Snatchers," fully displayed the continuity of the media scare with those historical fears
about bodily invasion, individual and national, that are often considered endemic to the paranoid style of American political culture.[3] Indeed, the rhetoric of computer culture, in common with the medical
discourse of AIDS research, has fallen in line with the paranoid, strategic mode of Defense Department rhetoric. Each language-repertoire

is obsessed with hostile threats to bodily and technological immune systems; every event is a ballistic maneuver in the game of microbiological war, where the governing metaphors are indiscriminately drawn from cellular genetics and cybernetics alike. As a counterpoint to the tongue-in-cheek artificial intelligence (AI) tradition of seeing humans as "information-exchanging environments," the imagined life of computers has taken on an organicist shape, now that they too are subject to cybernetic "sickness" or disease. So too the development of interrelated systems, such as Internet itself, has further added to the structural picture of an interdependent organism, whose component members, however autonomous, are all nonetheless affected by the "health" of each individual constituent. The growing interest among scientists in developing computer programs that will simulate the genetic behavior of living organisms (in which binary numbers act like genes) points to a future where the border between organic and artificial life is less and less distinct.

In keeping with the increasing use of biologically derived language to describe mutations in systems theory, conscious attempts to link the AIDS crisis with the information security crisis have pointed out that both kinds of virus, biological and electronic, take over the host cell/program and clone their carrier genetic codes by instructing the hosts to make replicas of the viruses. Neither kind of virus, however, can replicate itself independently; they are pieces of code that attach themselves to other cells/programs—just as biological viruses need a host cell, computer viruses require a host program to activate them. The Internet virus was not, in fact, a virus, but a worm, a program that can run independently and therefore *appears* to have a life of its own. The worm replicates a full version of itself in programs and systems as it moves from one to another, masquerading as a legitimate user by guessing the user passwords of locked accounts. Because of this autonomous existence, the worm can be seen to behave as if it were an organism with some kind of purpose or teleology, and yet it has none. Its only "purpose" is to reproduce and infect. If the worm has no inbuilt antireplication code, or if the code is faulty, as was the case with the Internet worm, it will make already-infected computers repeatedly accept further replicas of itself, until their memories are clogged. A much quieter worm than that engineered by Morris would have moved more slowly, as one supposes a "worm" should, protecting itself from detection by ever more subtle camouflage, and propagating its cumulative effect of operative systems inertia over a much longer period of time.

In offering such descriptions, however, we must be wary of attributing a teleology/intentionality to worms and viruses that can be ascribed only, and, in most instances, speculatively, to their authors. There is no reason a cybernetic "worm" might be expected to behave in any fundamental way like a biological worm. So, too, the assumed intentionality of its author distinguishes the human-made cybernetic virus from the case of the biological virus, the effects of which are fated to be received and discussed in a language saturated with human-made structures and narratives of meaning and teleological purpose. Writing about the folkloric theologies of significance and explanatory justice (usually involving retribution) that have sprung up around the AIDS crisis, Judith Williamson has pointed to the radical implications of this collision between an intentionless virus and a meaning-filled culture:

> Nothing could be more meaningless than a virus. It has no point, no purpose, no plan; it is part of no scheme, carries no inherent significance. And yet nothing is harder for us to confront than the complete absence of meaning. By its very definition, meaninglessness cannot be articulated within our social language, which is a system *of* meaning: impossible to include, as an absence, it is also impossible to exclude—for meaninglessness isn't just the opposite of meaning, it is the end of meaning, and threatens the fragile structures by which we make sense of the world.[4]

No such judgment about meaninglessness applies to the computer security crisis. In contrast to HIV's lack of meaning or intentionality, the meaning of cybernetic viruses is always already replete with social significance. This meaning is related, first of all, to the author's local intention or motivation, whether psychic or fully social, whether wrought out of a mood of vengeance, a show of bravado or technical expertise, a commitment to a political act, or in anticipation of the profits that often accrue from the victims' need to buy an antidote from the author. Beyond these local intentions, however, which are usually obscure or, as in the Morris case, quite inscrutable, there is an entire set of social and historical narratives that surround and are part of the "meaning" of the virus: the coded anarchist history of the youth hacker subculture; the militaristic environments of search-and-destroy warfare (a virus has two components—a carrier and a "warhead"), which, because of the historical development of computer technology, constitute the family values of information technoculture; the experimental research environments in which creative designers are encouraged to work; and the conflictual history of pure and applied ethics in the science and technology communities—to name just a few. A similar list could be drawn up to explain the widespread and varied *response* to computer viruses, from the amused concern of the cognoscenti to the hysteria of the casual user, and from the research community and the manufacturing industry to the morally aroused legislature and the mediated culture at large. Every one of these explanations and narratives is the result of social and cultural processes and values; consequently, there is very little about the virus itself that is "meaningless." Viruses can no more be seen as an objective, or necessary, result of the "objective" development of technological systems than technology in general can be seen as an objective, determining agent of social change.

For the sake of polemical economy, I would note that the cumulative effect of all the viral hysteria has been twofold. First, it has resulted in a windfall for software producers, now that users' blithe disregard for makers' copyright privileges has eroded in the face of the security panic. Used to fighting halfhearted rearguard actions against widespread piracy practices, or reluctantly acceding to buyers' desire for software unencumbered by top-heavy security features, software vendors are now profiting from the new public distrust of program copies. So too the explosion in security consciousness has hyperstimulated the already fast growing sectors of the security system industry and the data encryption industry. In line with the new imperative for everything from "vaccinated" workstations to "sterilized" networks, it has created a brand-new market of viral vaccine vendors who will sell you the virus (a one-time only immunization shot) along with its antidote—with names like Flu Shot +, ViruSafe, Vaccinate, Disk Defender, Certus, Viral Alarm, Antidote, Virus Buster, Gatekeeper, Ongard, and Interferon. Few of the antidotes are very reliable, however, especially since they pose an irresistible intellectual challenge to hackers who can easily rewrite them in the form of ever more powerful viruses. Moreover, most corporate managers of computer systems and networks know that by far the great majority of their intentional security losses are a result of insider sabotage and monkeywrenching.

In short, the effects of the viruses have been a profitable clamping down on copyright delinquency and the generation of the need for entirely new industrial production of viral suppressors to contain the fallout. In this respect, it is easy to see that the appearance of viruses could hardly, in the long run, have benefited industry producers more. In the same vein, the networks that have been hardest hit by the security squeeze are not restricted-access military or corporate systems, but networks like Internet, set up on trust to facilitate the open academic exchange of data, information, and research, and watched over by its sponsor, DARPA. It has not escaped the notice of conspiracy theorists that the military intelligence community, obsessed with "electronic warfare," actually stood to learn a lot from the Internet virus; the virus effectively "pulsed the system," exposing the sociological behavior of the system in a crisis situation.[5]

The second effect of the virus crisis has been more overtly ideological. Virus-conscious fear and loathing have clearly fed into the paranoid climate of privatization that increasingly defines social identities in the new post-Fordist order. The result—a psychosocial closing of the ranks around fortified private spheres—runs directly counter to the ethic that we might think of as residing at the architectural heart of information technology. In its basic assembly structure, information technology is a technology of processing, copying, replication, and simulation, and therefore does not recognize the concept of private information property. What is now under threat is the rationality of a shareware culture, ushered in as the achievement of the hacker counterculture that pioneered the personal computer revolution in the early seventies against the grain of corporate planning.

There is another story to tell, however, about the emergence of the virus scare as a profitable ideological moment, and it is the story of how teenage hacking has come to be defined increasingly as a potential threat to normative educational ethics and national security alike. The story of the creation of this "social menace" is central to the ongoing attempts to rewrite property law in order to contain the effects of the new information technologies that, because of their blindness to the copyrighting of intellectual property, have transformed the way in which modern power is exercised and maintained. Consequently, a deviant social class or group has been defined and categorized as "enemies of the state" to help rationalize a general law-and-order clampdown on free and open information exchange. Teenage hackers' homes are now habitually raided by sheriffs and FBI agents using strong-arm tactics, and jail sentences are becoming a common punishment. Operation Sundevil, a nationwide Secret Service operation conducted in the spring of 1990, involving hundreds of agents in fourteen cities, is the most recently publicized of the hacker raids that have produced several arrests and seizures of thousands of disks and address lists in the last two years.[6]

In one of the many harshly punitive prosecutions against hackers in recent years, a judge went so far as to describe "bulletin boards" as "hi-tech street gangs." The editors of *2600*, the magazine that publishes information about system entry and exploration indispensable to the hacking community, have pointed out that any single invasive act, such as trespass, that involves the use of computers is considered today to be infinitely more criminal than a similar act undertaken without computers.[7] To use computers to execute pranks, raids, fraud, or theft is to incur automatically the full repressive wrath of judges urged on by the moral panic created around hacking feats over the last two decades. Indeed, there is a strong body of pressure groups pushing for new criminal legislation that will define "crimes with computers" as a special category of crime, deserving "extraordinary" sentences and punitive measures. Over that same space of time, the term *hacker* has lost its semantic link with the journalistic *hack*, suggesting a professional toiler who uses unorthodox methods. So too its increasingly criminal connotation today has displaced the more innocuous, amateur mischief-maker-cum-media-star role reserved for hackers until a few years ago.

In response to the gathering vigor of this "war on hackers," the most common defenses of hacking can be presented on a spectrum that runs from the appeasement or accommodation of corporate interests to drawing up blueprints for cultural revolution. (a) Hacking performs a benign industrial service of uncovering security deficiencies and design flaws. (b) Hacking, as an experimental, free-form research activity, has been responsible for many of the most progressive developments in software development. (c) Hacking, when not purely recreational, is an elite educational practice that reflects the ways in which the development of high technology has outpaced orthodox forms of institutional education. (d) Hacking is an important form of watchdog counterresponse to the use of surveillance technology and data gathering by the state, and to the increasingly mono-lithic communications power of giant corporations. (e) Hacking, as guerrilla know-how, is essential to the task of maintaining fronts of cultural resistance and stocks of oppositional knowledge as a hedge against a technofascist future. With all of these and other arguments in mind, it is easy to see how the social and cultural *management* of hacker activities has become a complex process that involves state policy and legislation at the highest levels. In this respect, the virus scare has become an especially convenient vehicle for obtaining public and popular consent for new legislative measures and new powers of investigation for the FBI.[8]

Consequently, certain celebrity hackers have been quick to play down the zeal with which they pursued their earlier hacking feats, while reinforcing the *deviant* category of "technological hooligan-ism" reserved by moralizing pundits for "dark-side" hacking. Hugo Cornwall, British author of the best-selling *Hacker's Handbook,* pre-sents a Little England view of the hacker as a harmless fresh-air en-thusiast who "visits advanced computers as a polite country rambler might walk across picturesque fields." The owners of these proper-ties are like "farmers who don't mind careful ramblers." Cornwall notes that "lovers of fresh-air walks obey the Country Code, involving such items as closing gates behind one and avoiding damage to crops and livestock" and suggests that a similar code ought to "guide your rambles into other people's computers; the safest thing to do is sim-ply browse, enjoy and learn." By contrast, any rambler who "ven-tured across a field guarded by barbed wire and dotted with notices warning about the Official Secrets Act would deserve most that hap-pened thereafter." [9] Cornwall's quaint perspective on hacking has a certain "native charm," but some might think that this beguiling pic-ture of patchwork-quilt fields and benign gentlemen farmers glosses over the long bloody history of power exercised through feudal and postfeudal land economy in England, while it is barely suggestive of the new fiefdoms, transnational estates, dependencies, and principal-ities carved out of today's global information order by vast corpora-tions capable of bypassing the laws and territorial borders of sover-eign nation-states. In general, this analogy with "trespass" laws, which compares hacking to breaking and entering other people's homes, restricts the debate to questions about privacy, property, pos-sessive individualism, and, at best, the excesses of state surveillance, while it closes off any examination of the activities of the corporate owners and institutional sponsors of information technology (the almost exclusive "target" of most hackers).[10]

Cornwall himself has joined the lucrative ranks of ex-hackers who either work for computer security firms or write books about security for the eyes of worried corporate managers.[11] A different, though related, genre is that of the penitent hacker's "confession," produced for an audience thrilled by tales of high-stakes adventure at the keyboard, but written in the form of a computer security handbook. The best example of the "I Was a Teenage Hacker" genre is Bill (aka "The Cracker") Landreth's *Out of the Inner Circle: The True Story of a Computer Intruder Capable of Cracking the Nation's Most Secure Computer Systems,* a book about "people who can't 'just say no' to computers." In full complicity with the deviant picture of the hacker as "public enemy," Landreth recirculates every official and media cliché about subversive conspiratorial elites by recounting the putative exploits of a high-level hackers' guild called the Inner Circle. The author himself is presented in the book as a former keyboard junkie who now praises the law for having made a good moral example of him:

> If you are wondering what I am like, I can tell you the same things I told the judge in federal court: Although it may not seem like it, I am pretty much a normal American teenager. I don't drink, smoke or take drugs. I don't steal, assault people, or vandalize property. The only way in which I am really different from most people is in my fascination with the ways and means of learning about computers that don't belong to me.[12]

Sentenced in 1984 to three years' probation, during which time he was obliged to finish his high school education and go to college, Landreth concludes: "I think the sentence is very fair, and I already know what my major will be." As an aberrant sequel to the book's contrite conclusion, however, Landreth vanished in 1986, violating his probation, only to face later a stiff five-year jail sentence — a sorry victim, no doubt, of the recent crackdown.

Cyber-Counterculture?

At the core of Steven Levy's best-seller *Hackers* (1984) is the argument that the hacker ethic, first articulated in the 1950s among the famous MIT students who developed multiple-access user systems, is libertarian and crypto-anarchist in its right-to-know principles and its advocacy of decentralized technology. This hacker ethic, which has remained the preserve of a youth culture for the most part, asserts the basic right of users to free access to all information. It is a principled attempt, in other words, to challenge the tendency to use technology to form information elites. Consequently, hacker activities were presented in the eighties as a romantic countercultural tendency, celebrated by critical journalists like John Markoff of the *New York Times,* by Stewart Brand of *Whole Earth Catalog* fame, and by New Age gurus like Timothy Leary in the flamboyant *Reality Hackers.* Fueled by sensational stories about phone phreaks like Joe Egressia (the blind eight-year-old who discovered the tone signal of the phone company by whistling) and Captain Crunch, groups like the Milwaukee 414s, the Los Angeles ARPAnet hackers, the SPAN Data Travellers, the Chaos Computer Club of Hamburg, the British Prestel hackers, *2600*'s BBS, "The Private Sector," and others, the dominant media representation of the hacker came to be that of the "rebel with a modem," to use Markoff's term, at least until the more recent "war on hackers" began to shape media coverage.

On the one hand, this popular folk hero persona offered the romantic high profile of a maverick though nerdy cowboy whose fearless raids upon an impersonal "system" were perceived as a welcome tonic in the gray age of technocratic routine. On the other hand, he was something of a juvenile technodelinquent who hadn't yet learned the difference between right and wrong—a wayward figure whose technical brilliance and proficiency differentiated him nonetheless from, say, the maladjusted working-class J.D. street-corner boy of the 1950s (hacker mythology, for the most part, has been almost exclusively white, masculine, and middle-class). One result of this media profile was a persistent infantilization of the hacker ethic—a way of trivializing its embryonic politics, however finally complicit with dominant technocratic imperatives or with entrepreneurial-libertarian ideology one perceives these politics to be. The second result was to reinforce, in the initial absence of coercive jail sentences, the high educational stakes of training the new technocratic elites to be responsible in their use of technology. Never, the given wisdom goes, has a creative elite of the future been so in need of the virtues of a liberal education steeped in Western ethics!

The full force of this lesson in computer ethics can be found laid out in the official Cornell University report on the Robert Morris affair. Members of the university commission set up to investigate the affair make it quite clear in their report that they recognize the student's academic brilliance. His hacking, moreover, is described as a "juvenile act" that had no "malicious intent" but that amounted, like plagiarism—the traditional academic heresy—to a dishonest transgression of other users' rights. (In recent years, the privacy movement within the information community—a movement mounted by liberals to protect civil rights against state gathering of information—has actually been taken up and used as a means of criminalizing hacker activities.) As for the consequences of this juvenile act, the report proposes an analogy that, in comparison with Cornwall's *mature* English country rambler, is thoroughly American, suburban, middle-class, and *juvenile*. Unleashing the Internet worm was like "the driving of a golf-cart on a rainy day through most houses in the neighborhood. The driver may have navigated carefully and broken no china, but it should have been obvious to the driver that the mud on the tires would soil the carpets and that the owners would later have to clean up the mess." [13]

In what stands out as a stiff reprimand for his alma mater, the report regrets that Morris was educated in an "ambivalent atmosphere" where he "received no clear guidance" about ethics from "his peers or mentors" (he went to Harvard!). But it reserves its loftiest academic contempt for the press, whose heroization of hackers has been so irresponsible, in the commission's opinion, as to cause even further damage to the standards of the computing profession; media exaggerations of the courage and technical sophistication of hackers "obscures the far more accomplished work of students who complete their graduate studies without public fanfare," and "who subject their work to the close scrutiny and evaluation of their peers, and not to the interpretations of the popular press." [14] In other words, this was an inside affair, to be assessed and judged by fellow professionals within an institution that reinforces its authority by means of internally self-regulating codes of professionalist ethics, but rarely addresses its ethical relationship to society as a whole (acceptance of defense grants and the like). Generally speaking, the report affirms the genteel liberal ideal that professionals should not need laws, rules, procedural guidelines, or fixed guarantees of safe and responsible conduct. Apprentice professionals ought to have acquired a good conscience by osmosis from a liberal education rather than from some specially prescribed course in ethics and technology.

The widespread attention commanded by the Cornell report (attention from the Association of Computing Machinery, among others) demonstrates the industry's interest in how the academy invokes liberal ethics in order to assist in the managing of the organization of the new specialized knowledge about information technology. Despite or, perhaps, because of the report's steadfast pledge to the virtues and ideals of a liberal education, it bears all the marks of a legitimation crisis inside (and outside) the academy surrounding the new and all-important category of computer professionalism. The increasingly specialized design knowledge demanded of computer professionals means that codes that go beyond the old professionalist separation of mental and practical skills are needed to manage the division that a hacker's functional talents call into question, between a purely mental pursuit and the pragmatic sphere of implementing knowledge in the real world. "Hacking" must then be designated as a strictly *amateur* practice; the tension, in hacking, between *interestedness* and *disinterestedness* is different from, and deficient in relation to, the proper balance demanded by professionalism. Alternately, hacking can be seen as the amateur flip side of the professional ideal—a disinterested love in the service of interested parties and institutions. In either case, it serves as an example of professionalism gone wrong, but not very wrong.

In common with the two responses to the virus scare described earlier—the profitable reaction of the computer industry and the self-empowering response of the legislature—the Cornell report shows how the academy uses a case like the Morris affair to strengthen its own sense of moral and cultural authority in the sphere of professionalism, particularly through its scornful indifference to and aloofness from the codes and judgments exercised by the media—its diabolic competitor in the field of knowledge. Indeed, for all the trumpeting about excesses of power and disrespect for the law of the land, the revival of ethics, in the business and science disciplines in the Ivy League and on Capitol Hill (both awash with ethical fervor in the post-Boesky and post-Reagan years), is little more than a weak liberal response to working flaws or adaptational lapses in the social logic of technocracy.

To complete the scenario of morality play example-making, however, we must also consider that Morris's father was chief scientist of the National Computer Security Center, the National Security Agency's public effort at safeguarding computer security. A brilliant programmer and code breaker in his own right, he had testified in Washington in 1983 about the need to deglamorize teenage hacking, comparing it to "stealing a car for the purpose of joyriding." In a further Oedipal irony, Morris Sr. may have been one of the inventors, while at Bell Labs in the 1950s, of a computer game involving self-perpetuating programs that were a prototype of today's worms and viruses. Called Darwin, its principles were incorporated, in the 1980s, into a popular hacker game called Core War, in which autonomous "killer" programs fought each other to the death.[15]

With the appearance, in the Morris affair, of a patricidal object who is also the Pentagon's guardian angel, we now have many of the classic components of countercultural cross-generational conflict. What I want to consider, however, is how and where this scenario differs from the definitive contours of such conflicts that we recognize as having been established in the sixties; how the Cornell hacker Morris's relation to, say, campus "occupations" today is different from that evoked by the famous image of armed black students emerging from a sit-in on the Cornell campus; how the relation to technological ethics differs from Andrew Kopkind's famous statement, "Morality begins at the end of a gun barrel," which accompanied the publication of the "do-it-yourself Molotov cocktail" design on the cover of a 1968 issue of the *New York Review of Books;* or how hackers' prized potential access to the networks of military systems warfare differs from the prodigious Yippie feat of levitating the Pentagon building. It may be that, like the J.D. rebel without a cause of the fifties, the disaffiliated student dropout of the sixties, and the negationist punk of the seventies, the hacker of the eighties has come to serve as a visible public example of moral maladjustment, a hegemonic test case for redefining the dominant ethics in an advanced technocratic society. (Hence the need for each of these deviant figures to come in different versions—lumpen, radical chic, and Hollywood-style.)

What concerns me here, however, are the different conditions that exist today for recognizing countercultural expression and activism.

Twenty years later, the technology of hacking and viral guerrilla warfare occupies a similar place in countercultural fantasy as the Molotov cocktail design once did. While such comparisons are not particularly sound, I do think they conveniently mark a shift in the relation of countercultural activity to technology, a shift in which a software-based technoculture, organized around outlawed libertarian principles about free access to information and communication, has come to replace a dissenting culture organized around the demonizing of abject hardware structures. Much, though not all, of the sixties counterculture was formed around what I have elsewhere called the *technology of folklore*—an expressive congeries of preindustrialist, agrarianist, Orientalist, antitechnological ideas, values, and social structures. By contrast, the cybernetic countercultures of the nineties are already being formed around the *folklore of technology*—mythical feats of survivalism and resistance in a data-rich world of virtual environments and posthuman bodies—which is where many of the SF- and technology-conscious youth cultures have been assembling in recent years.[16]

There is no doubt that this scenario makes countercultural activity more difficult to recognize and therefore to define as politically significant. It was much easier, in the sixties, to *identify* the salient features and symbolic power of a romantic preindustrialist cultural politics in an advanced technological society, especially when the destructive evidence of America's supertechnological invasion of Vietnam was being daily paraded in front of the public eye. However, in a society whose techno-political infrastructure depends increasingly upon greater surveillance and where foreign wars are seen through the lens of laser-guided smart bombs, cybernetic activism necessarily relies on a much more covert politics of identity, since access to closed systems requires discretion and dissimulation. Access to digital systems still requires only the authentification of a signature or pseudonym, not the identification of a real surveillable person, so there exists a crucial operative gap between authentication and identification. (As security systems move toward authenticating access through biological signatures—the biometric recording and measurement of physical characteristics such as palm or retinal prints, or vein patterns on the backs of hands—the hackers' staple method of systems entry through purloined passwords will be further challenged.) By the same token, cybernetic identity is never used up—it can be re-created, reassigned, and reconstructed with any number of different names and under different user accounts. Most hacks, or technocrimes, go unnoticed or unreported for fear of publicizing the vulnerability of corporate security systems, especially when the hacks are performed by disgruntled employees taking their vengeance on management. So too authoritative identification of any individual hacker, whenever it occurs, is often the result of accidental leads rather than systematic detection. For example, Captain Midnight, the video pirate who commandeered a satellite a few years ago to interrupt broadcast TV viewing, was traced only because a member of the public reported a suspicious conversation heard over a crossed telephone line.

Eschewing its core constituency among white males of the pre-professional-managerial class, the hacker community may be expanding its parameters outward. Hacking, for example, has become a feature of young adult novel genres for girls.[17] The elitist class profile of the hacker prodigy as that of an undersocialized college nerd has become democratized and customized in recent years; it is no longer exclusively associated with institutionally acquired college expertise, and increasingly it dresses streetwise. In a recent article that documents the spread of the computer underground from college whiz kids to a broader youth subculture termed "cyberpunks," after the movement among SF novelists, the original hacker phone phreak Captain Crunch is described as lamenting the fact that the cyberculture is no longer an "elite" one, and that hacker-valid information is much easier to obtain these days.[18]

For the most part, however, the self-defined hacker underground, like many other protocountercultural tendencies, has been restricted to a privileged social milieu, further magnetized by the self-understanding of its members that they are the apprentice architects of a future dominated by knowledge, expertise, and "smartness," whether human or digital. Consequently, it is clear that the hacker cyberculture is not a dropout culture; its disaffiliation from a domestic parent culture is often manifest in activities that answer, directly or indirectly, to the legitimate needs of industrial R&D. For example, this hacker culture celebrates high productivity, maverick forms of creative work energy, and an obsessive identification with on-line endurance (and endorphin highs)—all qualities that are valorized by the entrepreneurial codes of silicon futurism. In a critique of the myth of the hacker-as-rebel, Dennis Hayes debunks the political romance woven around the teenage hacker:

> They are typically white, upper-middle-class adolescents who have taken over the home computer (bought, subsidized, or tolerated by parents in the hope of cultivating computer literacy). Few are politically motivated although many express contempt for the "bureaucracies" that hamper their electronic journeys. Nearly all demand unfettered access to intricate and intriguing computer networks. In this, teenage hackers resemble an alienated shopping culture deprived of purchasing opportunities more than a terrorist network.[19]

While welcoming the sobriety of Hayes's critique, I am less willing to accept its assumptions about the political implications of hacker activities. Studies of youth subcultures (including those of a privileged middle-class formation) have taught us that the political meaning of certain forms of cultural "resistance" is notoriously difficult to read. These meanings are either highly coded or expressed indirectly through media—private peer languages, customized consumer styles, unorthodox leisure patterns, categories of insider knowledge and behavior—that have no fixed or inherent political significance. If cultural studies of this sort have proved anything, it is that the often symbolic, not wholly articulate, expressivity of a youth culture can seldom be translated directly into an articulate political philosophy. The significance of these cultures lies in their embryonic or *protopolitical* languages and technologies of opposition to dominant or parent systems of rules. If hackers lack a "cause," then they are certainly not the first youth culture to be characterized in this dismissive way. In particular, the left has suffered from the lack of a cultural politics capable of recognizing the power of cultural expressions that do not wear a mature political commitment on their sleeves.

So too the escalation of activism-in-the-professions in the last two decades has shown that it is a mistake to condemn the hacker impulse on account of its class constituency alone. To cede the "ability to know" on the grounds that elite groups will enjoy unjustly privileged access to technocratic knowledge is to cede too much of the future. Is it of no political significance at all that hackers' primary fantasies often involve the official computer systems of the police, armed forces, and defense and intelligence agencies? And that the rationale for their fantasies is unfailingly presented in the form of a defense of civil liberties against the threat of centralized intelligence and military activities? Or is all of this merely a symptom of an apprentice elite's fledgling will to masculine power? The activities of the Chinese student elite in the prodemocracy movement have shown that unforeseen shifts in the political climate can produce startling new configurations of power and resistance. After Tiananmen Square, Party leaders found it imprudent to purge those high-tech engineer and computer cadres who alone could guarantee the future of any planned modernization program. On the other hand, the authorities rested uneasy knowing that each cadre (among the most activist groups in the student movement) is a potential hacker who can have the run of the communications house if and when he or she wants.

On the other hand, I do agree with Hayes's perception that the media have pursued their romance with the hacker at the cost of underreporting the much greater challenge posed to corporate employers by their employees. It is in the arena of conflicts between workers and management that most high-tech "sabotage" takes place. In the mainstream everyday life of office workers, mostly female, there is a widespread culture of unorganized sabotage that accounts for infinitely more computer downtime and information loss every year than is caused by destructive, "dark-side" hacking by celebrity cybernetic intruders. The sabotage, time theft, and strategic monkeywrenching deployed by office workers in their engineered electromagnetic attacks on data storage and operating systems might range from the planting of time or logic bombs to the discrete use of electromagnetic Tesla coils or simple bodily friction: "Good old static electricity discharged from the fingertips probably accounts for close to half the disks and computers wiped out or down every year." [20] More skilled operators, intent on evening a score with management, often utilize sophisticated hacking techniques. In many cases, a coherent networking culture exists among female console operators, where, among other things, tips about strategies for slowing down the temporality of the work regime are circulated. While these threats from below are fully recognized in their boardrooms, corporations dependent upon digital business machines are obviously unwilling to advertise how acutely vulnerable they actually are to this kind of sabotage. It is easy to imagine how organized computer activism could hold such companies for ransom. As Hayes points out, however, it is more difficult to mobilize any kind of labor movement organized upon such premises:

> Many are prepared to publicly oppose the countless dark legacies of the computer age: "electronic sweatshops," military technology, employee surveillance, genotoxic water, and ozone depletion. Among those currently leading the opposition, however, it is apparently deemed "irresponsible" to recommend an active computerized resistance as a source of worker's power because it is perceived as a medium of employee crime and "terrorism." [21]

Processed World, the "magazine with a bad attitude," with which Hayes has been associated, is at the forefront of debating and circulating these questions among office workers, regularly tapping into the resentments borne out in on-the-job resistance.

While only a small number of computer users would recognize and include themselves under the label of "hacker," there are good reasons for extending the restricted definition of *hacking* down and across the caste hierarchy of systems analysts, designers, programmers, and operators to include all high-tech workers, no matter how inexpert, who can interrupt, upset, and redirect the smooth flow of structured communications that dictates their positions in the social networks of exchange and determines the pace of their work schedules. To put it in these terms, however, is not to offer any universal definition of hacker agency. There are many social agents, for example, in job locations who are dependent upon the hope of technological *reskilling,* and for whom sabotage or disruption of communicative rationality is of little use; for such people, definitions of hacking that are reconstructive, rather than deconstructive, are more appropriate. A good example is the crucial role of worker technoliteracy in the struggle of labor against automation and deskilling. When worker education classes in computer programming were discontinued by management at the Ford Rouge plant in Dearborn, Michigan, United Auto Workers members began to publish a newsletter called the *Amateur Computerist* to fill the gap. [22] Among the columnists and correspondents in the magazine have been veterans of the Flint sit-down strikes who see a clear historical continuity between the problem of labor organization in the thirties and the problem of automation and deskilling today. Workers' computer literacy is seen as essential not only to the demystification of the computer and the reskilling of workers, but also to labor's capacity to intervene in decisions about new technologies that might result in shorter hours and thus in "work efficiency" rather than worker efficiency.

The three social locations I have mentioned above all express different class relations to technology: the location of an apprentice technical elite, conventionally associated with the term *hacking*; the location of the female high-tech office worker, involved in "sabotage"; and the location of the shop-floor worker, whose future depends on technological reskilling. All therefore exhibit different ways of *claiming back* time dictated and appropriated by technological processes, and of establishing some form of independent control over the work relation so determined by the new technologies. All, then, fall under a broad understanding of the politics involved in any extended description of hacker activities.

The Culture and Technology Question

Faced with these proliferating practices in the workplace, on the teenage cult fringe, and increasingly in mainstream entertainment, where, over the last five years, the cyberpunk sensibility in popular fiction, film, and television has caught the romance of the popular taste for the outlaw technology of human/machine interfaces, we are obliged, I think, to ask old kinds of questions about the new silicon order that the evangelists of information technology have been deliriously proclaiming for more than twenty years. The postindustrialists' picture of a world of freedom and abundance projects a bright millenarian future devoid of work drudgery and ecological degradation. This sunny social order, cybernetically wired up, is presented as an advanced evolutionary phase of society in accord with Enlightenment ideals of progress and rationality. By contrast, critics of this idealism see only a frightening advance in the technologies of social control, whose owners and sponsors are efficiently shaping a society, as Kevin Robins and Frank Webster put it, of "slaves without Athens" that is actually the inverse of the "Athens without slaves" promised by the silicon positivists.[23]

It is clear that one of the political features of the new post-Fordist order—economically marked by short-run production, diverse taste markets, flexible specialization, and product differentiation—is that the postindustrialists have managed to appropriate not only the utopian language and values of the alternative technology movements but also the Marxist discourse of the "withering away of the state" and the more compassionate vision of local, decentralized communications first espoused by the anarchist. It must be recognized that these are very popular themes and visions (advanced most publicly by Alvin Toffler and the neoliberal Atari Democrats, though also by leftist thinkers such as André Gortz, Rudolf Bahro, and Alain Touraine)—much more popular, for example, than the tradition of centralized technocratic planning espoused by the left under the Fordist model of mass production and consumption.[24] Against the postindustrialists' millenarian picture of a postscarcity harmony, in which citizens enjoy decentralized access to free-flowing information, it is necessary, however, to emphasize how and where actually existing cybernetic capitalism presents a gross caricature of such a postscarcity society.

One of the stories told by the critical left about new cultural technologies is that of monolithic, panoptical social control, effortlessly achieved through a smooth, endlessly interlocking system of networks of surveillance. In this narrative, information technology is seen as the most despotic mode of domination yet, generating not just a revolution in capitalist production but also a revolution in living—"social Taylorism"—that touches all cultural and social spheres in the home and in the workplace.[25] Through routine gathering of information about transactions, consumer preferences, and creditworthiness, a harvest of information about any individual's whereabouts and movements, tastes, desires, contacts, friends, associates, and patterns of work and recreation becomes available in the form of dossiers sold on the tradable information market, or is endlessly convertible into other forms of intelligence through computer matching. Advanced pattern recognition technologies facilitate the process of surveillance, while data encryption protects it from public accountability.[26]

While the debate about privacy has triggered public consciousness about these excesses, the liberal discourse about ethics and damage control in which that debate has been conducted falls short of the more comprehensive analysis of social control and social management offered by left political economists. According to one Marxist analysis, information is seen as a new kind of commodity resource that marks a break with past modes of production and that is becoming the essential site of capital accumulation in the world economy. What happens, then, in the process by which information, gathered up by data scavenging in the transactional sphere, is systematically converted into intelligence? A surplus value is created for use elsewhere. This surplus information value is more than is needed for public surveillance; it is often information, or intelligence, culled from consumer polling or statistical analysis of transactional behavior, that has no immediate use in the process of routine public surveillance. Indeed, it is this surplus bureaucratic capital that is used for the purpose of forecasting social futures, and consequently applied to the task of managing the behavior of mass or aggregate units within those social futures. This surplus intelligence becomes the basis of a whole new industry of futures research that relies upon computer technology to simulate and forecast the shape, activity, and behavior of complex social systems. The result is a possible system of social management that far transcends the questions about surveillance that have been at the discursive center of the privacy debate.[27]

To challenge further the idealists' vision of postindustrial light and magic, we need only look inside the semiconductor workplace itself, which is home to the most toxic chemicals known to man (and woman, especially since women of color often make up the majority of the microelectronics labor force), and where worker illness is measured not in quantities of blood spilled on the shop floor but in the less visible forms of chromosome damage, shrunken testicles, miscarriages, premature deliveries, and severe birth defects. Semiconductor workers exhibit an occupational illness rate that by the late seventies was already three times higher than that of manufacturing workers, at least until the federal rules for recognizing and defining levels of injury were changed under the Reagan administration. Protection gear is designed to protect the product and the clean room from the workers, and not vice versa. Recently, immunological health problems have begun to appear that can be described only as a kind of chemically induced AIDS, rendering the T-cells dysfunctional rather than depleting them like virally induced AIDS.[28] In addition to the extraordinarily high stress patterns of VDT operators in corporate offices, the use of keystroke software to monitor and pace office workers has become a routine part of job performance evaluation programs. Some 70 percent of corporations use electronic surveillance or other forms of quantitative monitoring of their workers. Every bodily movement can be checked and measured, especially trips to the toilet. Federal deregulation has meant that the limits of employee work space have shrunk, in some government offices, below that required by law for a two-hundred-pound laboratory pig.[29] Critics of the labor process seem to have sound reasons to believe that rationalization and quantification are at last entering their most primitive phase.

These, then, are some of the features of that critical left position—or what is sometimes referred to as the "paranoid" position—on information technology, which imagines or constructs a totalizing, monolithic picture of systematic domination. While this story is often characterized as conspiracy theory, its targets—technorationality, bureaucratic capitalism—are usually too abstract to fit the picture of a social order planned and shaped by a small, conspiring group of centralized power elites.

Although I believe that this story, when told inside and outside the classroom, for example, is an indispensable form of "consciousness-raising," it is not always the best story to tell. While I am not comfortable with the "paranoid" labeling, I would argue that such narratives do little to discourage paranoia. The critical habit of finding unrelieved domination everywhere has certain consequences, one of which is to create a siege mentality, reinforcing the inertia, helplessness, and despair that such critiques set out to oppose in the first place. What follows is a politics that can speak only from a victim's position. And when knowledge about surveillance is presented as systematic and infallible, self-censoring is sure to follow. In the psychosocial climate of fear and phobia aroused by the virus scare, there is a responsibility not to be alarmist or to be scared, especially when, as I have argued, such moments are profitably seized upon by the sponsors of control technology. In short, the picture of a seamlessly panoptical network of surveillance may be the result of a rather undemocratic, not to mention unsocialist, way of thinking, predicated upon the recognition of people solely as victims. It is redolent of the old sociological models of mass society and mass culture, which cast the majority of society as passive and lobotomized in the face of the cultural patterns of modernization. To emphasize, as Robins and Webster and others have done, the power of the new technologies to transform despotically the "rhythm, texture, and experience" of everyday life, and meet with no resistance in doing so, is not only to cleave, finally, to an epistemology of technological determinism, but also to dismiss the capacity of people to make their own use of new technologies.[30]

The seamless "interlocking" of public and private networks of information and intelligence is not as smooth and even as the critical school of hard domination would suggest. In any case, compulsive gathering of information is no *guarantee* that any interpretive sense will be made of the files or dossiers, while some would argue that the increasingly covert nature of surveillance is a sign that the "campaign" for social control is not going well. One of the most pervasive popular arguments against the panoptical intentions of the masters of technology is that their systems do not work. Every successful hack or computer crime in some way reinforces the popular perception that information systems are not infallible. And the announcements of military-industrial spokespersons that the fully automated battlefield is on its way run up against an accumulated stock of popular skepticism about the operative capacity of weapons systems. These misgivings are born of decades of distrust for the plans and intentions of the military-industrial complex, and were quite evident in the widespread cynicism about the Strategic Defense Initiative. The military communications system, for example, worked so poorly and so farcically during the U.S. invasion of Grenada that commanders had to call each other on pay phones: ever since then, the command-and-control code of ARPAnet technocrats has been C^5—Command, Control, Communication, Computers, and Confusion.[31] Even in the Gulf War, which has seen the most concerted effort on the part of the military-industrial-media complex to suppress evidence of such technical dysfunctions, the Pentagon's vaunted information system has proved no more, and often less, resourceful than the mental agility of its analysts.

I am not suggesting that alternatives can be forged simply by encouraging disbelief in the infallibility of existing technologies (pointing to examples of the appropriation of technologies for radical uses, of course, always provides more visibly satisfying evidence of empowerment), but technoskepticism, while not a *sufficient* condition for social change, is a *necessary* condition. Stocks of popular technoskepticism are crucial to the task of eroding the legitimacy of those cultural values that prepare the way for new technological developments: values and principles such as the inevitability of material progress, the "emancipatory" domination of nature, the innovative autonomy of machines, the efficiency codes of pragmatism, and the linear juggernaut of liberal Enlightenment rationality—all increasingly under close critical scrutiny as a wave of environmental consciousness sweeps through the electorates of the West. Technologies do not shape or determine such values. These values already preexist the technologies, and the fact that they have become deeply embodied in the structure of popular needs and desires then provides the green light for the acceptance of certain kinds of technology. The principal rationale for introducing new technologies is that they answer to already-existing intentions and demands that may be perceived as "subjective" but that are never actually within the control of any single set of conspiring individuals. As Marike Finlay has argued, just as technology is possible only in given discursive situations, one of which is the desire of people to have it for reasons of empowerment, so capitalism is merely the site, and not the source, of the power that is often autonomously attributed to the owners and sponsors of technology.[32]

There is no frame of technological inevitability that has not already interacted with popular needs and desires, no introduction of new machineries of control that has not already been negotiated to some degree in the arena of popular consent. Thus the power to design architecture that incorporates different values must arise from the popular perception that existing technologies are not the only ones, nor are they the best when it comes to individual and collective empowerment. It was this kind of perception—formed around the distrust of big, impersonal, "closed" hardware systems, and the desire for small, decentralized, interactive machines to facilitate interpersonal communication—that "built" the PC out of hacking expertise in the early seventies. These were as much the partial "intentions" behind the development of microcomputing technology as deskilling, monitoring, and information gathering are the intentions behind the corporate use of that technology today. The growth of public data networks, bulletin board systems, alternative information and media links, and the increasing cheapness of desktop publishing, satellite equipment, and international data bases are as much the result of local political "intentions" as the fortified net of globally linked, restricted-access information systems is the intentional fantasy of those who seek to profit from centralized control. The picture that emerges from this mapping of intentions is not an inevitably technofascist one, but rather the uneven result of cultural struggles over values and meanings.

It is in this respect—in the struggle over values and meanings— that the work of cultural criticism takes on its special significance as a full participant in the debate about technology. Cultural criticism is already fully implicated in that debate, if only because the culture and education industries are rapidly becoming integrated within the vast information service conglomerates. The media we study, the media we publish in, and the media we teach within are increasingly part of the same tradable information sector. So too, our common intellectual discourse has been significantly affected by the recent debates about postmodernism (or culture in a postindustrial world) in which the euphoric, addictive thrill of the technological sublime has figured quite prominently. The high-speed technological fascination that is characteristic of the postmodern condition can be read, on the one hand, as a celebratory capitulation on the part of intellectuals to the new information technocultures. On the other hand, this celebratory strain attests to the persuasive affect associated with the new cultural technologies, to their capacity (more powerful than that of their sponsors and promoters) to generate pleasure and gratification and to win the struggle for intellectual as well as popular consent.

Another reason for the involvement of cultural critics in the technology debates has to do with our special critical knowledge of the way in which cultural meanings are produced—our knowledge about the politics of consumption and what is often called the politics of representation. This is the knowledge that demonstrates that there are limits to the capacity of productive forces to shape and determine consciousness. It is a knowledge that insists on the ideological or interpretive dimension of technology as a culture that can and must be used and consumed in a variety of ways that are not reducible to the intentions of any single source or producer, and whose meanings cannot simply be read off as evidence of faultless social reproduction. It is a knowledge, in short, that refuses to add to the "hard domination" picture of disenfranchised individuals watched over by some scheming panoptical intelligence. Far from being understood solely as the concrete hardware of electronically sophisticated objects, technology must be seen as a lived, interpretive practice for people in their everyday lives. To redefine the shape and form of that practice is to help create the need for new kinds of hardware and software.

One of the aims of this essay has been to describe and suggest a wider set of activities and social locations than is normally associated with the practice of hacking. If there is a challenge here for cultural critics, then it might be presented as the obligation to make our knowledge about technoculture into something like a hacker's knowledge, capable of penetrating existing systems of rationality that might otherwise be seen as infallible; a hacker's knowledge, capable of reskilling, and therefore of rewriting the cultural programs and reprogramming the social values that make room for new technologies; a hacker's knowledge, capable also of generating new popular romances around the alternative uses of human ingenuity. If we are to take up that challenge, we cannot afford to give up what technoliteracy we have acquired in deference to the vulgar faith that tells us it is always acquired in complicity, and is thus contaminated by the poison of instrumental rationality, or because we hear, often from the same quarters, that acquired technological competence simply glorifies the inhuman work ethic. Technoliteracy, for us, is the challenge to make a historical opportunity out of a historical necessity.

NOTES

1. Bryan Kocher, "A Hygiene Lesson," *Communications of the ACM*, 32 (January 1989), 3.

2. Jon A. Rochlis and Mark W. Eichen, "With Microscope and Tweezers: The Worm from MIT's Perspective," *Communications of the ACM*, 32 (June 1989), 697.

3. Philip Elmer-DeWitt, "Invasion of the Body Snatchers," *Time* (September 26, 1988), 62–67.

4. Judith Williamson, "Every Virus Tells a Story: The Meaning of HIV and AIDS," *Taking Liberties: AIDS and Cultural Politics*, ed. Erica Carter and Simon Watney (London: Serpent's Tail/ICA, 1989), 69.

5. "Pulsing the system" is a well-known intelligence process in which, for example, planes deliberately fly over enemy radar installations in order to determine what frequencies they use and how they are arranged. It has been suggested that Morris Sr. and Morris Jr. worked in collusion as part of an NSA operation to pulse the Internet system, and to generate public support for a legal clampdown on hacking. See Allan Lundell, *Virus! The Secret World of Computer Invaders That Breed and Destroy* (Chicago: Contemporary Books, 1989), 12–18. As is the case with all such conspiracy theories, no actual conspiracy need have existed for the consequences—in this case, the benefits for the intelligence community—to have been more or less the same.

6. For details of these raids, see *2600: The Hacker's Quarterly,* 7 (Spring 1990), #7.

7. "Hackers in Jail," *2600: The Hacker's Quarterly,* 6 (Spring 1989), 22–23. The recent Secret Service action that shut down *Phrack,* an electronic newsletter operating out of St. Louis, confirms *2600*'s thesis; a nonelectronic publication would not be censored in the same way.

8. This is not to say that the new laws cannot themselves be used to protect hacker institutions, however. *2600* has advised operators of bulletin boards to declare them private property, thereby guaranteeing protection under the Electronic Privacy Act against unauthorized entry by the FBI.

9. Hugo Cornwall, *The Hacker's Handbook,* 3rd ed. (London: Century, 1988), 181, 2–6. In Britian, for the most part, hacking is still looked upon as a matter for the civil, rather than the criminal, courts.

10. Discussions about civil liberties and property rights, for example, tend to preoccupy most of the participants in the electronic forum published as "Is Computer Hacking a Crime?" in *Harper's,* 280 (March 1990), 45–58.

11. See Hugo Cornwall, *Data Theft* (London: Heinemann, 1987).

12. Bill Landreth, *Out of the Inner Circle: The True Story of a Computer Intruder Capable of Cracking the Nation's Most Secure Computer Systems* (Redmond, Wash.: Tempus, Microsoft, 1989), 10.

13. *The Computer Worm: A Report to the Provost of Cornell University on an Investigation Conducted by the Commission of Preliminary Enquiry* (Ithaca, N.Y.: Cornell University, 1989).

14. Ibid., 8.

15. A. K. Dewdney, the "computer recreations" columnist at *Scientific American,* was the first to publicize the details of this game of battle programs in an article in the May 1984 issue of the magazine. In a follow-up article in March 1985, "A Core War Bestiary of Viruses, Worms, and Other Threats to Computer Memories," Dewdney described the wide range of "software creatures" that readers' responses had brought to light. A third column, in March 1989, was written, in an exculpatory mode, to refute any connection between his original advertisement of the Core War program and the spate of recent viruses.

16. Andrew Ross, *No Respect: Intellectuals and Popular Culture* (New York: Routledge, 1989), 212. Some would argue, however, that the ideas and values of the sixties counterculture were only fullfilled in groups like the People's Computer Company, which ran Community Memory in Berkeley, or the Homebrew Computer Club, which pioneered personal microcomputing. So too the Yippies had seen the need to form YIPL, the Youth International Party Line, devoted to "anarcho-technological" projects, which put out a newsletter called *TAP* (alternately the *Technological American Party* and the *Technological Assistance Program*). In its depoliticized form, which eschewed the kind of destructive "dark-side" hacking advocated in its earlier incarnation, *TAP* was eventually the progenitor of *2600.* A significant turning point, for example, was *TAP*'s decision not to publish plans for the hydrogen bomb (the *Progressive* did so)— bombs would destroy the phone system, which the *TAP* phone phreaks had an enthusiastic interest in maintaining.

17. See Alice Bach's "Phreakers" series, which narrates the mystery-and-suspense adventures of two teenage girls: *The Bully of Library Place* (New York: Dell, 1988),

Double Bucky Shanghai (New York: Dell, 1987), *Parrot Woman* (New York: Dell, 1987), *Ragwars* (New York: Dell, 1987), and others.

18. John Markoff, "Cyberpunks Seek Thrills in Computerized Mischief," *New York Times* (November 26, 1988), 1, 28.

19. Dennis Hayes, *Behind the Silicon Curtain: The Seductions of Work in a Lonely Era* (Boston: South End, 1989), 93. One striking historical precedent for the hacking subculture, suggested to me by Carolyn Marvin, was the widespread activity of amateur or "ham" wireless operators in the first two decades of the century. Initially lionized in the press as boy-inventor heroes for their technical ingenuity and daring adventures with the ether, this white middle-class subculture was increasingly demonized by the U.S. Navy (whose signals the amateurs prankishly interfered with), which was crusading for complete military control of the airwaves in the name of national security. The amateurs lobbied with democratic rhetoric for the public's right to access the airwaves, and, although partially successful in their case against the Navy, lost out ultimately to big commercial interests when Congress approved the creation of a broadcasting monopoly after World War I in the form of RCA. See Susan J. Douglas, *Inventing American Broadcasting 1899–1922* (Baltimore: Johns Hopkins University Press, 1987), 187–291.

20. "Sabotage," *Processed World,* 11 (Summer 1984), 37–38.

21. Hayes, *Behind the Silicon Curtain,* 99.

22. *The Amateur Computerist,* available from R. Hauben, P.O. Box 4344, Dearborn, MI 48126.

23. Kevin Robins and Frank Webster, "Athens without Slaves . . . or Slaves without Athens? The Neurosis of Technology," *Science as Culture,* 3 (1988), 7–53.

24. See Boris Frankel, *The Post-Industrial Utopians* (Oxford: Basil Blackwell, 1987).

25. See, for example, the collection of essays edited by Vincent Mosco and Janet Wasko, *The Political Economy of Information* (Madison: University of Wisconsin Press, 1988), and Dan Schiller, *The Information Commodity* (Oxford University Press, forthcoming).

26. Tom Athanasiou and Staff, "Encryption and the Dossier Society," *Processed World,* 16 (1986), 12–17.

27. Kevin Wilson, *Technologies of Control: The New Interactive Media for the Home* (Madison: University of Wisconsin Press, 1988), 121–25.

28. Hayes, *Behind the Silicon Curtain,* 63–80.

29. "Our Friend the VDT," *Processed World,* 22 (Summer 1988), 24–25.

30. See Kevin Robins and Frank Webster, "Cybernetic Capitalism," *The Political Economy of Information,* ed. Vincent Mosco and Janet Wasko (Madison: University of Wisconsin Press, 1988), 44–75.

31. Barbara Garson, *The Electronic Sweatshop* (New York: Simon & Schuster, 1988), 244–45.

32. See Marike Finlay's Foucauldian analysis, *Powermatics: A Discursive Critique of New Technology* (London: Routledge & Kegan Paul, 1987). A more conventional culturalist argument can be found in Stephen Hill, *The Tragedy of Technology* (London: Pluto, 1988).

Brownian Motion:
Women, Tactics, and Technology
Constance Penley

Near the end of *Star Trek V: The Final Frontier*, Captain Kirk, thought to be dead but rescued finally by Spock and some exceptionally helpful Klingons, stands facing his first officer on the bridge of the Klingon ship. Glad to be alive, he moves toward Spock and reaches for him with both hands. Spock interrupts the embrace with "Please, Captain, not in front of the Klingons." Kirk directs a brief glance toward the known universe's most macho aliens, then turns back to Spock to exchange a complicitous look before lowering his hands. Most members of the audience probably took this teasing one-liner as just another instance of what actor/director William Shatner has called the "tongue-in-cheek" campiness of the original TV series,[1] a quality he is clearly trying to reintroduce into this, perhaps the last, of the *Star Trek* films with the original cast. But for a small minority of the audience, a group of female fans who have for years dedicated themselves to writing and publishing underground pornographic stories about Kirk and Spock as spacefaring lovers, this scene came as a delightful surprise, an astonishing recognition of their desires by a *Star Trek* industry that has up until now met those desires with curt dismissals, cute evasions, or disdainful silence.[2] It is not yet clear what it will mean for the fans to have had their desires recognized, their fantasies ratified, not only by William Shatner but, indirectly, the Great Bird of the Galaxy, the fans' pet name for producer Gene Roddenberry, who created *Star Trek* and still has creative control over all *Star Trek* productions. It may be that the fandom will see some erosion of the pleasure that comes from its tightly secret, marginalized solidarity, which has been built on the fans' delight in how

By permission of the artist.

their guerrilla erotics have shocked and enraged (and, surely, sometimes amused) the producers of *Star Trek,* fans involved in official
Star Trek fandom, and "mundanes" (as they call nonfans) who may

have stumbled across some of the steamier stories in fanzines like *Naked Times, Off Duty, Fever,* and *Final Frontier.* What is more likely, however, is that this fleeting public recognition of their hitherto illicit desires will only spur them on, insofar as their solidarity as a group also rests on their pride in having created both a unique, hybridized genre that ingeniously blends romance, pornography, and utopian science fiction and a comfortable yet stimulating social space in which women can manipulate the products of mass-produced culture to stage a popular debate around issues of technology, fantasy, and everyday life. This, of course, is my version of it.[3] The fans would say they are just having fun.

Women have been writing *Star Trek* pornography since at least 1976, mostly in the United States, but also in Britain, Canada, and Australia. The idea did not begin with one person who then spread it to others, but seems to have arisen spontaneously in various places beginning in the mid-seventies, as fans recognized, through seeing the episodes countless times in syndication and on their own taped copies, that there was an erotic homosexual subtext there, or at least one that could easily be *made* to be there. Most of the writers and readers started off in "regular" *Star Trek* fandom, and many are still involved in it, even while they pursue their myriad activities in what is called "K/S" or "slash" fandom. The slash between K(irk) and S(pock) serves as a code to those purchasing by mail[4] amateur fanzines (or "zines") that the stories, poems, and artwork published there concern a same-sex relationship between the two men. Such a designation stands in contrast to "ST," for example, with no slash, which stands for action-adventure stories based on the *Star Trek* fictional universe, or "adult ST," which refers to stories containing sexual scenes, but heterosexual ones only, say between Captain Kirk and Lt. Uhura, or Spock and Nurse Chapel. Other media male couples have been "slashed" in the zines, like Starsky and Hutch (S/H), Simon and Simon (S/S), or *Miami Vice*'s Crockett and Tubbs (or Castillo) (M/V). The slash premise, however, seems to work best with science fiction couples; K/S was the first slash writing and still predominates, with its nearest rival being a newer science fiction fandom based on *Blakes' 7,* a British television show broadcast from 1978 to 1981. As we shall see, the popularity and success of SF slash are due to the range and complexity of discourses that are possible in a genre that could be described as romantic pornography radically shaped and reworked by the themes and tropes of science fiction.

The conventions of the science fiction genre seem to offer several important advantages to the writing of "pornography by women, for women, with love," as Joanna Russ described slash writing in a 1985 essay of that title.[5] It has been argued that science fiction, seemingly the most sexless of genres, is in fact engrossed with questions of sexual difference and sexual relations, which it repeatedly addresses alongside questions of other kinds of differences and relations: humans and aliens, humans and machines, time travelers and those they visit, and so on.[6] It has also been argued, most recently by Sarah Lefanu in *Feminism and Science Fiction,* that science fiction offers women writers a freedom not available in mainstream writing because its generic form, with its overlooked roots in the female gothic novel and nineteenth-century feminist utopian literature, permits a fusing of political concerns with the "playful creativity of the imagination."[7] And this is so, she says, even though science fiction has been historically a male preserve.

Lefanu limits herself, however, to rounding up the usual suspects, those women science fiction writers with more or less conscious feminist politics who have written stories and novels that have, nonetheless, been able to make it into the SF mainstream. (Although she notes the existence of K/S writing in her introduction, she does not discuss it in any detail.) These writers include such well-known figures as Ursula Le Guin, Joanna Russ, Suzy McKee Charnas, and James Tiptree, Jr. (Alice Sheldon), as well as feminist writers in the literary mainstream, like Marge Piercy and Margaret Atwood, who have on occasion made use of SF themes and tropes. The women writers I want to discuss, however, are amateur writers, few of whom would be willing to identify themselves as feminists, even though their writing and their fan activity might seem to offer an indirect (and sometimes not so indirect) commentary on issues usually seen as feminist, such as women's lack of social and economic equality, their having to manage a double-duty work and domestic life, and their being held to much greater standards of physical beauty than men. The tension here between the feminist concerns of the fans and their unwillingness to be seen as feminists can teach those who work in the field of women and popular culture a great deal about how political issues get articulated in everyday life outside the accepted languages of feminism and the left. What is at issue here is not discovering in this female fan culture a pre- or protopolitical language that could then be evaluated from the perspective of "authentic" feminist thought, but, rather, finding alternative and unexpected ways of thinking and speaking about women's relation to the new technologies of science, the body, and the mind.

Michel de Certeau uses the term "Brownian motion" to describe the tactical maneuvers of the relatively powerless when attempting to resist, negotiate, or transform the system and products of the relatively powerful.[8] He defines *tactics* as guerrilla actions involving hit-and-run acts of apparent randomness. Tactics are not designed primarily to help users take over the system but to seize every opportunity to turn to their own ends forces that systematically exclude or marginalize them. These tactics are also *a way of thinking* and "show the extent to which intelligence is inseparable from the everyday struggles and pleasures that it articulates."[9] The only "product" of such tactics is one that results from "making do" (*bricolage*)—the process of combining already-existing heterogeneous elements. It is not a synthesis that takes the form of an intellectual discourse about an object; the form of its making is its intelligence. The K/S fans, however, seem to go de Certeau's "ordinary man" one better. They are not just reading, viewing, or consuming in tactical ways that offer fleeting moments of resistance or pleasure while watching TV, scanning the tabloids, or selecting from the supermarket shelves (to use some of his examples). They are producing not just intermittent, cobbled-together acts, but real products (albeit ones taking off from already-existing heterogeneous elements)—zines, novels, artwork, videos—that (admiringly) mimic and mock those of the industry they are "borrowing" from while offering pleasures found lacking in the original products. K/S fandom more than illustrates de Certeau's claim that consumption is itself a form of production. A mini-industry, but one that necessarily makes no money (the only thing keeping it from copyright suits), it has its own apparatuses of advertising and publishing; juried prizes (K/Star, Surak, and Federation Class of Excellence Awards); stars (the top editors, writers, and artists, but also fans who have become celebrities); house organ, *On the Double*; annual meetings, featuring charity fund-raisers (e.g., an art auction to support an animal shelter); music videos (with scenes from *Star Trek* reedited for their "slash" meanings); brilliant built-in market research techniques (the consumers are the producers and vice versa, since almost all of the slash readers are also its writers); and, increasingly, the elements of a critical apparatus, with its own theorists and historians.[10] The fandom has achieved a form of vertical integration—control over every aspect of production, distribution, and consumption—that the trust-busted film industry is only just now being allowed to dream about again in this era of Reagan-Bush deregulation.

Although this fan publishing apparatus could not exist without the prior existence of the *Star Trek* industry, its relation to that industry cannot be described as a parasitic one. Parasites often injure their hosts, but slash fandom in no way seeks to harm or destroy the world of *Star Trek,* even the particularly unsatisfying version presented in the new television series, *Star Trek: The Next Generation,* whose perceived cold, high-tech surfaces, affectless characters, and lack of humor have so far made it relatively impervious to slashing. Rather, the fans want only to *use* the system imposed by the other, a practice that, as de Certeau describes it, "redistributes its space; it creates at least a certain play in that order, a space for maneuvers of unequal forces and for utopian points of reference." This is where, he says, the "opacity" of popular culture manifests itself:

> [a] dark rock that resists all assimilation. . . . the subtle, stubborn, resistant activity of groups which, since they lack their own space, have to get along in a network of already established forces and representations. People have to make do with what they have. . . . the tactical and joyful dexterity of the mastery of a technique.[11]

In many ways, however, slash fans do more than "make do"; they make. Not only have they remade the *Star Trek* fictional universe to their own desiring ends, they have achieved it by enthusiastically mimicking the technologies of mass-market cultural production, and by constantly debating their own relation, as women, to those technologies, through both the way they make decisions about how to use the technological resources available to them and the way they rewrite bodies and technologies in their utopian romances.

"Appropriate Technology" and Slash Zine Publishing

The term *appropriate technology* refers to both everyday uses of technology that are appropriate to the job at hand and the way users decide how and what to appropriate. To avoid becoming dependent on sources that extract too high a price, or to ensure that the technology will be available to everyone, one appropriates only what is needed. The slashers (their name for themselves) are constantly involved in negotiating appropriate levels of technology for use within the fandom. The emphasis is on keeping the technology accessible and democratic, although this turns out to be easier said than done. The general perception among fans is that media zine editors give more attention to the appearance of the zines than do SF literature zine editors, and that overall they look a lot slicker—laser-printed and xeroxed rather than mimeoed, for example. And it is true that the media zines, and especially the slash zines, look very good. They are beautifully produced, with glossy, illustrated covers; spiral, velo, or even perfect binding; color xerography on the cover or inside, laser-printed type, and intricate page borders.

However, given the high level of everyday technological skills the fans must have developed in their jobs as nurses, teachers, office workers, draftswomen, copy shop managers, and so on, what is striking is that the zines, as good as they look, are not as slick as they could be. In part, this may arise from an impulse to keep them looking *slightly* tacky to give them that illegitimate pornographic cast. One of the binding forces of slash fan culture is the shared delight in the visual shock value of the zines. Although zine publishers claim that they cannot afford heavy-duty plain envelopes for mailing, I have often suspected that an important element of this fandom's pleasure lies in the illicit thrill of receiving in the mail (to the stares and smiles of one's mail carrier, friends, family, or colleagues) a half-torn envelope revealing a particularly juicy drawing of Kirk and Spock, their naked bodies arranged in some near-impossible position. But it is also likely that the publishing technology is only semideveloped because deliberate decisions have been made to keep the technology "appropriate," unintimidating, accessible, and hence democratic.

Workshops on the "how-to" of zine publishing are offered at each convention, and the zine editors who run the workshops are generous with their advice, sharing what they have learned from experience. Several helpful pamphlets, also full of advice, with step-by-step instructions on how to edit and publish a zine, are available from fan editors. One zine editor/publisher who works at a copy center put out a brochure advertising its services in this way:

> I am sure you are aware of the increasingly high cost of copying and binding and the difficulty in finding a printer who is quality-minded, reliable, economical and gives the confidentiality that your publications deserve. If you have missed deadlines, encountered poor printing quality and/or disapproving counter personnel, then look no further for a remedy—CopyMat can save the day.

Occasionally, however, there will be a fan revolt against even this apparently easy and democratic access to new publishing technologies. Many of the women in *Professionals* fandom (based on a British secret service show), one of the strongest non-SF slash alternatives to K/S, have almost entirely eschewed the zine publishing process in a bid to become what is called a "circuit fandom." If a fan wants to read the latest *Professionals* stories, she sends a stamped, self-addressed envelope to a designated fan in Illinois, who sends out ten of the most recently received stories. Or, if she is a new fan, she can ask the fan who manages the circuit to pick out the "ten best" stories to send to her. The fan does her own photocopying and sends the originals back to Illinois. *Professionals* stories are not usually advertised, edited, or even "published," but are simply disseminated in the most basic way imaginable, among fans who say they are fed up with what they see as the technological hassles of zine publishing, its ridiculously high standards for copyediting and illustration, and its resultant overprofessionalization. They also object to the "difficult" personalities of some zine editors, by which they mean editors who are seen to publish only themselves and their friends or who are thought to "censor" certain kinds of stories.[12]

The issue of observing appropriate levels of technology is a contested one for the fans. Fan editors who have both the skills and access to good equipment and who are unable to resist the lure of the latest technology ("Take advantage of CopyMat's Canon Laser Color Copier") often wish to produce ever more sophisticated looking zines. On the other hand, fan readers often voice their doubts about this tendency toward more professional looking publications, saying that they feel the "look" of the zine is entirely secondary to the content of the stories and the quality of the writing. And fan writers object when they feel that editors are spending less time copyediting and proofreading their work than on soliciting work from fan artists and perfecting their graphics technology. Complicating this debate even further is the fact that there are no clear divisions among readers, writers, artists, and editor/publishers, and therefore no correspondingly clear conflicts among their respective interests. Almost all fan readers are also writers, many are editors of their own zines, and some are also artists; some of the most enduring and prolific editor/publishers also perform all the other roles. Conflicting impulses, then, about appropriate levels of technology can be harbored in a single person. But all of the fans, no matter how much some of them might feel pulled toward a greater "professionalism," still pay formal homage to the shared desire to keep the technology of the publishing apparatus within the reach of all.

Just as the fans feel split over their relation to the available high-technology of modern desktop publishing, they also have conflicting desires about amateur versus professional levels of writing. The fandom is, first and foremost, militant in its desire to maintain an unintimidating milieu in which women who want to write can do so without fear of being held to external, professional standards of "good writing." The pride and pleasure the fans take in their writing are immense. They talk about it as a form of escape from the pressures of daily domestic and work lives that is superior to what they see as the more passive escape provided by romance reading, for example. But they value it even more for the expression of individual creativity it allows. Above all, they recognize that they feel free to express themselves as writers only insofar as they can conceive of their writing as a hobby and nothing more. Even this commitment to thinking of what they do as a hobby, however, gets subtly subverted by the fans themselves. At the most basic level of standards of punctuation and typographical accuracy, for example, fans demand that writers and editors be meticulous in catching and correcting errors. Such errors, they claim, can break the erotic fantasy when they occur at important moments in the story. There are fan editors, of course, who defend themselves by saying that they would rather spend their time finding good stories than nitpicking over every typo and misspelling, but in general everyone prefers errorless stories.

The strong pull toward "professionalization" is described by the fans in terms of getting "hooked" or "contaminated" by the writing and editing process. One fan writer and editor came up with what she calls "the virus theory of fandom":

reading = contact
writing = infection
editing = full-blown disease

The virus theory of fandom attempts to account for the tendency to become fascinated and then obsessed with the craft of writing, to want to delve ever deeper into its techniques to produce something that pleases both the author and the readers. As two fan editors remarked during a slash convention writing panel, "Fans go in because they need to create something and then feel good when it goes out and others like it." Another possible motive for wanting to write, and write well, was voiced by a third editor on this panel: "It also makes you feel that you're not abnormal for having picked up on the relationship between Starsky and Hutch!"

The most palpable tension between the commitment to amateurism and the wish to perfect one's craft along more professional lines can be seen in the popular writing panel discussions and workshops offered at every slash convention. Although the fans want to learn all they can from the more experienced writers leading the discussion, they often tend to resist workshop leaders' emphasis on craft and technique in favor of a focus on inspiration and the "magic" of writing. And in response to a long series of very specific suggestions by one experienced fan writer about how to fashion a story idea as effectively as possible, another fan objected by insisting (albeit rather plaintively), "There's a place for stuff that's just so-so." But perhaps the greatest source of tension lies in the fans' knowledge of how many of their cohorts have "crossed over" into professional writing, many of whom (indeed, most of whom) maintain their relation to fandom because they still want to be part of that supportive community, and feel very loyal to it, even when they have been quite successful in commercial writing.

Many of the fans show visible pride in fellow slash writers who have gone pro, even those who have done so by deslashing and heterosexualizing their own or others' work to turn it into commercial stories or novels (some of the *Star Trek* paperback novels are based on slash stories or were written by slash writers). But their ambivalence, which is, I think, finally productive, still manifests itself. At a slash convention panel discussion on precisely this topic, one fan commented, "It's not our best writers who've gone pro." Another fan scoffed, "That's what we like to think!" I call it a productive ambivalence because it is one that impels the fans to debate not simply the merits of "amateur" versus "pro" writing. They must also address the assumptions shaping those categories, and do so by challenging the idea that only those who are already "credentialed" may be allowed access to the means of acquiring cultural capital through writing fiction or poetry.

The fans seem less concerned about their relation to video technologies than their relation to writing technologies. Although the video contest is often the high point of a slash convention, going on for an immensely raucous and pleasurable three or four hours, fewer of the fans are involved in making videos, or songtapes, as they are called. One reason is obvious: the greater difficulty of access to video equipment, and especially editing equipment, than to desktop publishing and photocopying technologies, which are often available in the fan's own workplace and can be used even while on the job.

That the fans are concerned to make this technology more available, however, can be seen in the scheduling at fan conventions of workshops such as one I attended called "Song Tapes for the Masses." The workshop, organized by two fan video artists known for making songtapes that are not particularly slick but are highly effective and popular with the fans, offered the novice or would-be songtape producer advice that was both practical and aesthetic. The songtapes are, in fact, music videos made right at home with two VCRs, an audio cassette deck, and a stopwatch. The organizers of the workshop handed out a helpful chart that allows the songtape maker to write down each line of, usually, a rock song with a love theme, the duration of each phrase, and the duration of the video segment that will be matched with it. The video segments are taken from fans' private collections of the 78 (plus the pilot) *Star Trek* episodes, and also the five *Star Trek* films, which are also on tape, copied from video store rentals.

The video artist begins by cataloguing all the scenes of Kirk and Spock together, and then selects the ones that, when matched with the music, will bring out what the fans call the "slash premise," that is, that the two men are in love and sexually involved with each other. As I said before, the advice ranges from the practical ("Look at your video material without sound; the sound will confuse you"; "Use the show's own cuts, fades, dissolves, etc., since you can't do them yourself") to the aesthetic ("Watch MTV or VH1 and just see what they do"; "Don't be too literal, for example, having a line like 'rain is pouring down' and showing someone in the shower").

What was stressed throughout the workshop was keeping the production as cheap as possible ("Time the song segments with a $5.00 stopwatch") and the built-in advantages of low-tech ("Hi-fi isn't important since it will come out mono on most machines anyway"). However, as the workshop progressed, the organizers, who have been making songtapes for many years, could not resist telling us about some of the neat things they have learned to do with this basic equipment and demonstrating a few of their shortcuts through what is admittedly a tedious and intensively labor-consuming process. The discussion began to get increasingly more technical, involving editing processes that would require additional, and relatively more expensive, equipment. For example, they suggested that everyone get a video insert machine ("no rollback") and a machine with a flying advance head ("no jittery shots or rainbows"). And, finally, at the very end of the workshop: "Really, what you should do is buy an editing machine." They recommended the Panasonic lap editor from Radio Shack for $147.00, a price that would be out of the range of many of the fans just coming to songtape production. One came away from the workshop, however, not with the sense of having been intimidated, but with the feeling that one could adopt whatever level of technology one felt able to handle and could afford, so clear and explicit had the advice and directions been.

One piece of technology about which the fans have no ambivalence whatever is the VCR, which, along with the zine publishing apparatus, is the lifeblood of the fandom. The ubiquitous VCR allows fans to copy episodes for swapping or for closer examination of their slash possibilities, and provides the basic technology for producing songtapes. Fans are deeply invested in VCR technology because it is cheap, widely available, easy to use, and provides both escape and a chance to criticize the sexual status quo. As one beautifully embroidered sampler at a fan art auction put it, "The more I see of men, the more I love my VCR."

Slash Tactics: Technologies of Writing

Just as the slash fans are constantly debating and negotiating their relation to technology within the fandom, this same concern appears in the fictions of the stories they write. In this respect, the fans' task of writing stories involving technology has been made easier for them. For such an elaborately produced science fiction show (even though much of it looks hokey to us now), the original *Star Trek* had a curiously ambivalent relation to the representation of futuristic technology. Although the producers of the show consulted scientists, engineers, and technicians in their efforts to make the technology plausible, they decided finally to give only the barest and sketchiest of outlines, to keep the design of the ship and the various scientific, medical, and military instruments extremely basic and simple. Not only was this decision an economical one (for example, some of Dr. McCoy's medical instruments were made from saltshakers), simplicity helped to ensure that the technology would not almost immediately look dated. "Phasers," "tricorders," "communicators," "scanners," "photon torpedoes," and "warp drive" were therefore designed to reveal their functions without divulging anything about how they were actually supposed to work. Franz Joseph's *Star Fleet Technical Manual*, first published in 1975, promises on its cover that one will find inside "detailed schematics of Star Fleet equipment," "navigational charts and equipment," and "intersteller space/warp technology," but fails to deliver anything but exhaustive descriptions of the way the instruments *look*, saying nothing of their functioning.[13]

While some fans have felt compelled to flesh out the sketchy contours of *Star Trek* technology, these have mostly been men. For example, an ad in a recent issue of *Datazine* for a zine called *Sensor Readings*, edited by Bill Hupe, says that it "features articles on warp factor cubed theory (by Tim Farley), shuttlecraft landing approach methods (by Steven K. Dixon), [and] the electronic printing methods available to fanzine editors today (by Randall Landers, a former Kinko's manager)." This is not to say that male *Star Trek* fans, who are more likely to edit the nonfiction zines, invariably exhibit a more developed relation to high technology than do women fans, who form the great majority of fiction zine writers and publishers. Rather, what I want to emphasize is that the women *Star Trek* fans, especially the slash fans, have defined *technology* in a way that includes the technologies of the body, the mind, and everyday life. It is a notion of technology that sees everything in the world (and out of this world)

as interrelated and subject to influence by more utopian and imaginative desires than those embodied in existing technological hardware.

This is the way one of the most prolific K/S writers and publishers,
Alexis Fegan Black, suggests that aspiring slash writers approach the
question of technology:

> Perhaps most importantly, WRITE WHAT YOU KNOW! That isn't to
> say that you can't write about being on board the *Enterprise*
> because you obviously have never been there. But if you *are* going
> to write a story that deals in mechanical or electronic details with
> the workings of the *Enterprise,* do so convincingly. A good rule of
> thumb is that it's best *not* to use what sci-fi writers call "pseudo-
> science" (cursory explanations of something dredged up solely
> from imagination) unless absolutely necessary.

But Black tells slash writers to take heart:

> Fortunately, the K/S genre is one wherein technology can usually be
> kept to a minimum. And writing what you know should be on a
> more emotional level than a technological one in most cases.

She goes on to say that she herself has been able to take up subjects
like martial arts, cryogenics, and metaphysics because she has already researched those areas and thus the effort to come up with
convincing detail does not take away from attention to the "emotional" aspects of the story. Black describes her own writing as "technomysticism," and even though her work is more directly inspired
by New Age ideas than most slash writing, it is instructive for understanding the genre's relation to technology to see how this extremely
influential and prolific writer (she has written more novels than any
other slash writer and manages the largest slash press) folds descriptions of outer space into meditations on inner space.

Science fiction writing is usually broken down into two schools:
so-called hard SF (men and machines colonizing the galaxies) and
soft SF (work that extrapolates from ideas found in the human and
social sciences, like sociology, psychology, or anthropology, rather
than the "natural" or physical sciences). There are, of course, countless examples of blurred boundaries and crossovers (and this is increasingly so), but work like Black's would seem at first sight firmly
ensconced in the soft school. Her work, however, does not read like,
say, Ursula Le Guin's, the avatar of SF based in the human and social
sciences, and indeed its references are not to any disciplines recog-

nized by the academy, but to more popular ones derived from "pop" psychology, New Age ideas, environmental awareness, and a peculiarly American brand of libertarianism that believes itself to be the inheritor of Kennedy-era liberalism—equated by the fans with what they see as producer Gene Roddenberry's most important idea, the Vulcan philosophy of IDIC, Infinite Diversity in Infinite Combination.

A closer look at Black's most well-received novel, *Dreams of the Sleepers,* the first in a trilogy of fan award-winning books, will give a sense of what she means by "technomysticism." Published in 1985, *Dreams of the Sleepers* is a time-travel story, and, like most such stories, revels in the dizzying paradoxes of journeying through time. The zine/book begins with an editorial titled "What's It All About?" We are plunged into a narrative in which four men in black arrive at the author's home in vans with government license plates. Her home is next to a "missile testing range" that she is sure is really a government installation for detaining captured aliens. They want to know how she found out about Kirk and Spock, and say that they would like "to ask . . . a few questions about this manuscript. . . . What's this all about?" This little narrative turns into a proper editorial in which the author says that *Dreams of the Sleepers* aims to get the reader to ask the question, "What *would* this world be like without *Star Trek?*" We are then returned to the narrative, in which Alexis finds what she takes to be a prank letter under the door, a letter from Dr. McCoy to Admiral Nogura of Star Fleet Command. The letter accompanies a manuscript that McCoy claims he confiscated during the *Enterprise* crew's continuing mission into Earth's past, while they wait for Kirk and Spock's return, the two having inexplicably disappeared. Now the novel itself begins, with an entry in the captain's log: the crew has been ordered on a mission into Earth's past, around 1963, to be precise. Their mission is to find out everything they can about old Earth's early experimentation with "psychotronics," the psychic manipulation of reality. Meanwhile, Kirk and Spock are feeling the first stirrings of what they are slowly realizing is their love and passion for one another, although their relationship has not yet been consummated, and each man does not yet know the depth of the other man's feelings for him. Just before beaming down to Earth, Spock suggests that he and Kirk form a mind link "for security reasons" while on their mission (as a half-Vulcan, Spock has the ability to link up with another mind empathically and even telepathically). Through the

link, Kirk and Spock understand for the first time that their desire is
mutual.

Almost immediately after beaming down to the military/scientific
installation, which turns out to be the private but government-
funded Futura Technics, Kirk and Spock are captured and put into
life suspension units, but not before learning the purpose of the
project. Scientists have been lured to Futura Technics to work on life
suspension for space exploration, but soon find out that the project's
real aim is to harness the psychic energy of the "sleepers" for use as
defensive and offensive weapons, as well as sabotage of all kinds. Not
only humans have been captured and suspended, but also aliens. Indeed, it is a Klingon sleeper who will travel out of his body to carry
out the next scheduled mission, two days hence—the assassination
of John F. Kennedy! The head of the project explains to Kirk and
Spock, before putting them under, that certain people in the government and military fear Kennedy's popularity and believe that if he
lives, the country will become truly united and could then be led
into peace, not war. But they do not want peace to come so soon, and
certainly not on Kennedy's terms. They too want "peace," but only
after conquering the rest of the world with their psychotronic
weapons—the dreams of the sleepers—which will kill people by
making them *believe* they have been attacked by nuclear weapons.

The two men having been put to "sleep" in life suspension units, it
is Spock who first awakes into his astral form. Another astral traveler
who is also a sleeper in the complex teaches Spock how to move
around in space but also in time. He takes Spock twenty-two years
into the future to show him a world devastated by war and hints to
him that the end of the world is somehow linked to something Spock
and Kirk either did or did not do. Kirk finally awakes into his astral
body and joins Spock, eventually setting up housekeeping on the astral plane; finding a nicely decorated and uninhabited ranch house
nearby, they "move in" and begin to pass the time with elaborate sexual fantasies and lovemaking.

Spock takes Kirk forward twenty-two years to see the postapoca-
lyptic ruin of the planet. Kirk weeps for all the dead but also, more
selfishly, for himself, because if this future comes to pass, he will
never have been born, and he and Spock would never have been
able to come together as friends, lovers, or the twin souls they have
now become. Suddenly they notice that another man has material-

ized, sitting cross-legged under a tree. The man shakes hands with
Spock, then laughs with joy and disbelief. "I *knew* it! ... Damn! I
knew it.... You're *real.*" "Gene," the man under the tree, is, of
course, Gene Roddenberry, whom Spock immediately recognizes as
the key to changing the future: if this "strange messiah in polyester
leisure-wear" can only realize his dream for all to see—of a popu-
lous and peace-loving federation of all the galaxies' creatures—then
humans will be inspired to give up waging war on each other and go
to the stars instead.

To help him realize his dream, Spock links his mind with Gene's to
show him the future. He decides that it is the "logical" thing to do
even though Gene will also see his most private thoughts and will
understand the nature of his relationship with Kirk. But it is "a fair
price to purchase a world's survival." Gene promises that in return
he will find some way to help rescue them from the complex. Thus:
"On September 8, 1966 [the date the first episode of *Star Trek* was
broadcast], the future formed a tentative bond with the present, in-
terlinking its parts with the past. After three years, however, that link
was severed. But throughout the world, minds were altered in subtle
ways." Underachievers and autistic children begin functioning brilliantly. Technology is turned to peaceful purposes. Educational levels
rise dramatically and knowledge is no longer the property of the
elite. Advanced computers become available to everyone. Peace
breaks out all over. The space program expands, transforming science fiction into fact, and the first space shuttle is named *Enterprise*
by popular acclaim (this actually happened). Meanwhile, Kirk and
Spock dream on, not knowing about the new world they have helped
to create, and not knowing when or if their liberator will ever come.
They again travel into the future, but this time apocalypse has been
averted. Gene appears, saying maybe he's just an idealist, but he'd
like to think that they had had something to do with it, and shows
Kirk and Spock episodes of *Star Trek,* telling them how influential
the show has been, how many followers it has had, and so on.

In talking about *Star Trek* later, Kirk says to Spock that Gene was
wrong about one thing: " 'Space isn't the final frontier. ... *You* are!'
[The Vulcan replies,] 'Indeed. Then perhaps, Jim,' he suggested, lean-
ing closer to whisper softly into one ear, 'we should ... boldly go ...
where no man has gone before.' " This erotic exchange is mapped
onto the realization that they *must* return to the future so that they

can have existed to be able to go back into the past to make sure that *Star Trek* gets produced, the world gets saved, and humans go into space. (Kirk has a hard time following this, but fortunately the more intellectual Spock grasps the intricacies and paradoxes of time travel.) The novel ends with their dramatic rescue by the female security officer of the *Enterprise*.

The popularity of Black's *Dreams of the Sleepers* and its two sequels lies in the way she is able to elaborate her idea of "technomysticism" to express the deepest wish of *Star Trek* fan culture: that the fandom matter, that what the fans do can affect the world in significant ways. However, it is not enough for the critic to identify this wish and be satisfied with designating it as a *symptom* (of the fans' need, for example, for an imaginary family or community, or as a substitute for their lack of real social agency or cultural capital), which is precisely the way fan culture is usually discussed.[14] The conceptual strength of slash writing forces us to see that it is more interesting to look at what the fans are *doing* with this individually and collectively elaborated discourse than it is to discuss what it "represents." And, because this discourse is so imbued with utopian longings "to free the individual, through leisure, technology, and self-realization, to go out and meet others as equals instead of enemies," [15] it also begs for a reconsideration of the role and value of utopian thinking, especially when this form of popular argument is carried out in and through a mass-culture product, and by the relatively disempowered.

Because I am focusing on K/S, I am going to have less to say about how *Star Trek* fans in general use the fictional construct of *Star Trek* to articulate a *lived* relation to the world. Through their involvement in *Star Trek* fandom, fans take an enormous interest in the space program and are conscientious about following its development, the reasons for its failures, and the vicissitudes of its funding. Many fans think that their experience with extrapolatory fiction has given them a privileged sense of "thinking global" and that they are more likely than others to be concerned with environmental issues. They see themselves as firmly committed to a politics of equality and tolerance, devoted as they are to a rather sunny version of Kennedy-era liberalism.

So too the fans' anti-imperialist leanings, based on their respect for the Federation's Prime Directive of noninterference in other cultures, can be as shaky and inconsistent as they ever were on *Star Trek*.

Although fans delight in keeping a tally of the number of times the Prime Directive has been broken—so many times, in fact, that it has become the exception rather than the rule—their own tendency toward anti-imperialism is undermined in the same way the Prime Directive is undermined in the *Star Trek* universe: in the original show, as in its successor, it was never clear whether the *Enterprise* was on a scientific or a military mission. Furthermore, if the *Enterprise* is a United Federation of Planets starship, why is it called the *U.S.S. Enterprise?* In the original *Star Trek,* the new series, and the fandom itself, a preference for peaceful uses of technology and the principle of noninterference exist side by side with nationalistic patriotism and unrestrained affection for militarist uses of technology. A Trekker can thus, without apparent contradiction, adopt the precepts of IDIC, the Prime Directive, and the peaceful use of technology while still exulting in the U.S. "victory" in Grenada or enthusiastically supporting Star Wars/SDI. Slash writing often puts even more emphasis than does regular *Star Trek* fan writing on the crew's military mission because of the narrative need to provide an alibi for men being "forced" to come together for companionship, love, and even sex—the navy ship, along with the prison, the gym, the police precinct, and the wide-open range, is a conventional location in gay male pornography.

Trek fandom also shows its commitments in its social constituency: it is interracial, includes people of all ages, has a fair number of disabled members, is sexually balanced, and has a strong cross-class representation, though perhaps most members are in the pink-collar, "subprofessional," or high-tech service industry sectors. This is not to say that *Trek* fandom is incapable of self-contradiction, discrimination, bad politics, bad faith, and all the rest, but it *is* to maintain that this huge fandom (35,000 official members, with many thousands more in smaller, unofficial clubs, or unaffiliated) represents one of the most important popular sites for debating issues of the human and everyday relation to science and technology.

Slash fandom, however, as I have argued, extends these issues and debates about science and technology to the realm of minds and bodies. The K/Sers are constantly asking themselves why they are drawn to writing their sexual and social utopian romances across the bodies of two men, and why these two men in particular. The answers—and there are surely more than one—range from the pleasures of writing explicit same-sex erotica to the fact that writing a story about two men avoids the built-in inequality of the romance formula, in which dominance and submission are invariably the respective roles of male and female. There are also advantages to writing about a futuristic couple: it is far from incidental that women have chosen to write their erotic stories about a couple living in a fully automated world in which there will never be fights over who has to scrub the tub, take care of the kids, cook, or do the laundry. Indeed, one reason the fans give for their difficulty thus far in slashing *Star Trek: The Next Generation* is that children and families now live on the *Enterprise* (albeit in a detachable section!), and that those circumstances severely cut into the erotic possibilities.

All the same, one still wonders why these futuristic bodies—this couple of the twenty-third century—must be imagined and written as male bodies. Why are the women fans so alienated from their own bodies that they choose to write erotic fantasies only in relation to a nonfemale body? Some who have thought about this question, fans and critics alike, have tried to show that Kirk and Spock are not coded as male but are rather androgynous, even arguing that this was the case on the original show. Slash readers and writers would then be identifying with and eroticizing characters who combine traits of masculinity and femininity. However, the more I read of the slash literature, the more I am convinced that Kirk and Spock are clearly meant to be male. The first reason for this helps to answer the second question about the women fans' alienation from their own bodies. The bodies from which they are indeed alienated are twentieth-century women's bodies: bodies that are a legal, moral, and religious battleground, that are the site of contraceptive failure, that are publicly defined as *the* greatest potential danger to the fetuses they house, that are held to painfully greater standards of physical beauty than those of the other sex. Rejecting the perfect Amazons that populate much of female fantasy/sword-and-sorcery writing, the K/Sers opt instead for the project of at least *trying* to write real men. (From what I have seen and read in the fandom, I would argue that it is indeed a rejection of the Amazons' artificiality and not a rejection of lesbianism, even though most of the K/Sers are heterosexual.)

What must be remembered also is the K/Sers' penchant for "making do": when asked why they do not create original characters who could be women as well as men, they most often respond that they are just "working with what's out there," which happens to be the world of broadcast television, a world typically peopled with strong male characters with whom to identify and take as erotic objects. They also insist that one can enter the *Star Trek* world through the male characters only, since the female characters, like Uhura, Nurse Chapel, and Yeoman Rand, have always been quite marginal and un-satisfactory. The slashers' aggressive identification with the men and the taking of them as sexual objects, in fantasy scenarios in which the female fans can participate but are, finally, radically excluded, might also serve as an oblique commentary on the *homosociality* of our culture.

The desire to write real men can be carried out only within a project of "retooling" masculinity itself, which is precisely what K/S writing sets out to do. It is for this reason as well that Kirk and Spock must be clearly male and not mushily androgynous. This retooling is made easier by locating it in a science fiction universe that is both futuristic and offers several generic tropes that prove useful to the project. Feminists, as well as the fans in their daily lives, have had to confront the fact that we may not see the hoped-for "new" or "trans-formed" men in our lifetimes, and if the truth be told, we often rid-icule the efforts of men who try to remake themselves along feminist lines (as Donna Haraway says, "I'd rather go to bed with a cyborg than a sensitive man").[16] The idea of sexual equality, which will nec-essarily require a renovated masculinity, is taking a long time to be-come a lived reality and is hard to imagine, much less write. This dif-ficulty can be seen, for example, in the unsatisfying attempt to rewrite male romance characters in the new Silhouette Desire "Man of the Month" series, each volume of which features a male protag-onist trying to come to terms with his identity and his sexuality in a world that no longer gives clear messages about what will count as "masculinity," but still threatens dire consequences for those men who fail to attain it. The male characters in this series, so feebly sketched out, seem particularly unconvincing, and it is both painful and distasteful to have to share the man's consciousness and narra-tive point of view. More implausible yet is the heroine's passion, if only because it is so hard to believe that anyone would want these guys!

But Kirk and Spock, as rewritten by the slashers, are another mat-ter. If it has become difficult to imagine new men in the present day, then it may be easier to imagine them in a time yet to come. *Surely* three hundred years from now it will be better. Kirk and Spock *are* sensitive in the slash stories, as well as kind, strong, thoughtful, and humorous, but their being "sensitive" carries with it none of the associations of wimpiness or smug self-congratulation it does in the present day. Only in the future, it seems, will it be possible to conceive that yielding phallic power does not result in psychical castration or a demand to be praised extravagantly for having yielded that power. Kirk and Spock, however, are rarely written as already perfect; they too have to do some work on themselves. Although the characters are provided with the SF device of the Vulcan mind link, which allows them to communicate more intimately than today's men are thought to do, Kirk and Spock are typically shown learning to overcome the conditioning that prevents them from expressing their feelings. Spock, whose Vulcan training has led him to suppress his emotions totally, has to learn to accept his human or emotional side, since he is, after all, half human. And Kirk, raised an Iowa farm boy, must first recognize and then reject now-archaic ideas of masculinity that were the product of his extremely conventional upbringing. Many slash stories relegate "action" to the background to ensure the tightest possible focus on the two men undergoing this painful yet liberatory process of self-discovery and learning to communicate their feelings.

And although it is true that, by the fans' own admission, they usually heterosexualize Kirk and Spock's sexual practices,[17] often the major sign that Kirk and Spock are different from today's men is that they can freely discuss their own homosexual tendencies and learn not to be insulted or afraid if someone takes them for a gay couple. A feminist critic said to me once, "Isn't this slash stuff just fag-hagging?" Yes, there is an idealization of the gay male couple in this fan writing, one that I would argue is understandable, because it is a couple, after all, for whom love and work can be shared by two equals (a state of affairs the fans feel to be almost unattainable for a heterosexual couple). But there is also a comprehension of the fact that *all* men (and women) must be able to recognize their own homosexual tendencies if they are to have any hope of fundamentally changing oppressive sexual roles. The hateful term *fag-hagging* also obscures the very real appreciation the fans have of gay men in their efforts to re-

By permission of the artist.

define masculinity, and their feelings of solidarity with them insofar as gay men too inhabit bodies that are still a legal, moral, and religious battleground. This is a more likely explanation than fag-hagging (a term that insults everyone) for the fans' wish to offer a sympathetic portrayal of these two men who are lovers.

Slash does not stop with retooling the male psyche; it goes after the body as well. Some changes are cosmetic, others go deeper. Spock, for example, has extra erogenous zones (especially the tips of his pointed ears) and a double-ridged penis. But the greatest change concerns the plot device of *pon farr,* the heat suffered every seven years by all Vulcan males. The male goes into a blood fever *(plak tow),* can become very violent, and will die if he does not have sex, preferably with a mate. The slash fans are not making this up—in the thirty-fourth episode of *Star Trek* (written by Theodore Sturgeon), Spock goes into *pon farr,* begins to die, and is taken back to Vulcan by his comrades so that he can complete the mating ritual and live. *Pon farr* stories are so popular with the slash fans that a new zine called *Fever* has been started up to publish only *pon farr* stories. I think the fans relish these stories, in part, because they like the idea of men too being subject to a hormonal cycle, and indeed their version of Spock's pre-*pon farr* and *plak tow* symptoms are wickedly and humorously made to parallel those of PMS and menstruation, in a playful and transgressive leveling of the biological playing field. Another nice touch is that Kirk, because he is empathically bonded with the Vulcan through the mind link, does not have to be told when Spock is getting ready to go into *pon farr* or how he is feeling; in fact, he often shares Spock's symptoms.

But perhaps the most extreme retooling of the male body is seen in the stories in which Kirk and Spock have a baby. Few of these stories exist and they are generally badly reviewed by the fans, who feel that, even for them, the premise is too farfetched, and that, finally, pregnancy and child-rearing responsibilities get in the way of erotic fantasies. In one such story, Kirk and Spock are able to have a baby only after Dr. McCoy does a great deal of genetic engineering to create a fertilized ovum, and Scotty a great deal of mechanical and electronic engineering to build an exterior womb. Not only does it take four men to make a baby in this story (!), but the very awkwardness of the apparatus (at the level of story and discourse) and the fans' rejection of most Kirk-and-Spock-have-a-baby stories suggest that some feats of bodily technology, especially when they involve such substantial regendering, are still unimaginable and unwritable.

In slash fandom, then, and the writing practice that it supports, we find a bracing instance of the strength of the popular wish to think through and debate the issues of women's relation to the technologies of science, the mind, and the body, in both fiction and everyday life. I have argued that much can be learned from the way the slashers make individual and collective decisions about how they will use technology at home, at work, and at leisure, and how they creatively reimagine their world through making a tactics of technology itself.

NOTES

1. William Shatner, quoted in Allan Asherman, *The Star Trek Interview Book* (New York: Pocket Books, 1988), 19.

2. The producers and writers of *Star Trek,* as well as the principal actors, know of the existence of the fan fiction writing that depicts Kirk and Spock as lovers, and several have responded when pressed for their opinions of the fictional pairing (although none admits to having read any of the fan fiction). In his novelization of *Star Trek: The Motion Picture* (New York: Pocket Books, 1979), Gene Roddenberry, the head producer of *Star Trek* and its original creator, adds an editor's note to "report" Kirk's opinion (and Spock's, indirectly) on the matter. Here, Kirk is purportedly replying to the author's request for clarification:

> I was never aware of this *lovers* rumor, although I have been told that Spock encountered it several times. Apparently he had always dismissed it with his characteristic lifting of his right eyebrow which usually connoted some combination of surprise, disbelief, and/or annoyance. As for myself, although I have no moral or other objections to physical love in any of its many Earthly, alien, and mixed forms, I have always found my best gratification in that creature *woman*. Also, I would dislike being thought of as so foolish that I would select a love partner who came into sexual heat only once every seven years.

David Gerrold, science fiction and *Star Trek* writer who wrote *The World of Star Trek* (New York: Bluejay, 1984), takes a much harsher approach to trying to curtail fan desire to rewrite Kirk and Spock as lovers. He realized he could not write a book about *Star Trek* and the fan culture surrounding it without mentioning what he calls "the K/S ladies" (p. 197). He also felt he must say something, because even being told by Gene Roddenberry himself that Kirk and Spock are "just friends" has not stopped the K/S fans from projecting their own sexual fantasies onto *Star Trek,* which Gerrold insists is a nuisance to the producers and the fans. He claims that "more than one young would-be fan" has been prohibited from attending *Star Trek* conventions or reading the institutionally approved *Star Trek* magazines because his or her parents have seen this material. Pitting women against each other, he quotes one woman fan who objects to the fact that "too many" of the stories involve Kirk and Spock in sado-masochistic scenarios, which, she feels, doesn't reflect appropriate behavior for two of Star Fleet's finest officers. To round off his attack he quotes a gay male *Star Trek* fan who is offended by the depiction of gay men in the K/S writing. Finally, speaking for himself, he accuses the K/S fans of not having good manners because they have not followed the rules of Gene Roddenberry's "universe" (pp. 121–22).

Fans report that Leonard Nimoy, who plays Spock, has given the best response. Once, at a *Star Trek* convention, he replied to a request to give his opinion on the likelihood of Kirk and Spock being lovers by saying, "I don't know, I wasn't there."

3. "My version" is based on both my personal involvement in the fandom and my interest in the study of women and popular culture. I subscribe to many of the slash fanzines, attend conventions, and correspond with fans. I show my work on fandom to various fans, and they give me their comments on it. I do everything I can to protect the privacy of individual women in the fandom, but realize that writing about the fandom for a larger public may bring some unwanted attention. Since the fans' practices are so inherently interesting and useful to those concerned with the question of what women *do* (and *can do*) with the products of mass-produced culture—how they resist, negotiate, and even remake them—I hope the fans will not feel compromised by this use of their work. For a fuller discussion of the ethics, difficulties, and advantages of being in the position of both fan and feminist researcher, see my "Feminism, Psychoanalysis and the Study of Popular Culture," *Cultural Studies Now and in the Future,* ed. Lawrence Grossberg, Cary Nelson, and Paula Treichler (New York: Routledge, forthcoming).

4. The main source for ordering media fanzines is *Datazine,* P.O. Box 24590, Denver, CO 80224. A one-year subscription (six issues) is $12.00.

5. Joanna Russ, "Pornography by Women, for Women, with Love," *Magic Mommas, Trembling Sisters, Puritans and Perverts: Feminist Essays* (Trumansburg, N.Y.: Crossing, 1985).

6. See my "Time Travel, Primal Scene, and the Critical Dystopia," *The Future of an Illusion: Film, Feminism, and Psychoanalysis* (Minneapolis: University of Minnesota Press, 1989).

7. Sarah Lefanu, *Feminism and Science Fiction* (Bloomington: Indiana University Press, 1989), 2.

8. Michel de Certeau, *The Practice of Everyday Life,* trans. Steven F. Rendall (Berkeley: University of California Press, 1984).

9. Ibid., xx.

10. Those who have written on slash writing include Joanna Russ (see note 5); Patricia Frazer Lamb and Diana L. Veith, "Romantic Myth, Transcendence, and *Star Trek* zines," *Erotic Universe: Sexuality and Fantastic Literature* (New York: Greenwood, 1986); Henry Jenkins III, "*Star Trek,* Rerun, Reread, Rewritten: Fan Writing as Textual Poaching," *Critical Studies in Mass Communication,* 5 (June 1988); and Camille Bacon-Smith, "Spock among the Women," *New York Times Book Review* (November 16, 1986), 1, 26, 28.

11. de Certeau, *The Practice of Everyday Life,* 18. See Jenkins's extended use of de Certeau's notion of "poaching" to characterize fan writing behavior in "*Star Trek* Rerun, Reread, Rewritten."

12. Margaret Garrett has reminded me that there are other important reasons besides dissatisfaction with zine publishing for the fans' preferring to circulate *Professionals* stories in a less formal way. The writing of *Professionals* fan fiction began in England, and it is much more difficult there than in the United States to produce slash fiction discreetly. British fans do not have the same access to self-service copiers that U.S. fans have, and certainly could not afford to own copying machines to be used at home, as several U.S. slash editors do. *Professionals* stories thus were usually circulated in carbon copies, and only made it into photocopy form once they were sent to fans in the United States. The "circuit" started as a way to disseminate the unpublished British stories in this country. As the *Professionals* reading audience has grown, by leaps and bounds, both in North America and Australia, these stories are increasingly

being published in zines by American fans with their cheaper technology. Fans still support the "circuit," however, according to Garrett, for its freedom, availability, and low cost.

13. This lack of explicitness is, however, given a fictional motivation: the information in the manual, which was accidentally transmitted during a space-warp from the *Enterprise* computer to a computer at a military installation in Omaha, Nebraska, is incomplete; Star Fleet Command has deleted any technology not known to twentieth-century Earth technology, in order to preserve the Prime Directive's dictum of nonintervention in other cultures, including the culture of the Federation's own past.

14. See the first section of Jenkins, "*Star Trek* Rerun, Reread, Rewritten," for an overview of how fans are typically characterized by both academic and journalistic writers.

15. Kirk's thoughts, in Syn Ferguson's epic slash novel, *Courts of Honor* (1985).

16. Quoted in Constance Penley and Andrew Ross, "Cyborgs at Large: Interview with Donna Haraway," this volume.

17. Joanna Russ, "Pornography by Women," 83, points out the fans' tendency to heterosexualize Kirk and Spock's sexual practices. In my essay in *Cultural Studies Now and in the Future* (see note 3), I try to say why many slash fans want Kirk and Spock to be, however improbably, heterosexual.

"Penguin in Bondage": A Graphic Tale of Japanese Comic Books
Sandra Buckley

The last page of the August 1989 issue of *Penguin Club,* a Japanese pornographic comic book (*pooruno manga*), carries an advertisement for a computer software package titled "Penguin in Bondage." The title may seem more comic than pornographic to the English reader, but the description of the content of the computer program is far from funny. The player has to overcome a cast of some eighty monsters created by a mad scientist in order to survive the game. He (the readership is predominantly male) also has at his disposal an array of "bunnygirls" and "angels" and a set of "items" that can be used on either the monsters or the female characters. The player can also *kuroonu* (clone) his own females and program for a range of variables including level of complexity of game, preferred sexual activity, hormone levels, degree of intelligence, level of violence, and level of endurance of pain. The written script of the woman's "dialogue" with the player and the corresponding images appear on the screen in accordance with the story programmed by the player. The advertisement coins a new expression for this software—"YPG" or *yarashii pureeingu geemu.* The word *yarashii* could be translated as "obscene," but is usually used interchangeably in the comic books (*manga*) with the Jap-lish *erotikku* (erotic). This gives the direct translation of "erotic playing game."

YPG is only one of a rapidly increasing number of pornographic software games available in bookstores and by mail order. Far more violent games are available on the black market. These also feature do-it-yourself sexual violence games and come in a variety of "flavors," including sadomasochism, bestiality, rape, and pedophilia.

Advertisement for "Penguin in Bondage" software.

That Japan is at the forefront of pornographic software is not surprising. Nor is it surprising that a good deal of the software is being developed and marketed around the more popular of the pornographic comic books. In order to understand the "naturalness" of this progression, it is necessary to go back to the Edo period, when commercialized pornography first flourished, and then to move on to the late 1960s, when Japan entered a second wave of pornography. To write this history is to write a history of recuperation and containment.

The Japanese word for the comic books is *manga*. It is said to have been first coined by the Edo woodblock artist Hokusai Katsushika (1760–1849) to describe the comic woodblock images that were so popular at that time.[1] Even the better known and more successful woodblock artists such as Hokusai were not above producing pornographic *manga* prints for a rapidly expanding market. This was a quick and easy way to make money. The introduction and rapid refinement of woodblock printing during the Edo period arguably provided the most significant catalyst in the circulation of pornographic images at that time. The commercialization of pornography developed hand in hand with the technology of reproduction. The *shunga* are today generally referred to as Japanese erotic art, and yet they too date back to this same period and were printed as a form of poster art available for purchase by the Yoshiwara patrons. The *shunga* prints represented the whole range of sexual options available to the patron of the quarters—heterosexual, gay, lesbian, and masturbatory images show a vast repertoire of sexual practices for sale or fantasy. The patron who could not afford the "real thing"—the price tag on a high-ranking or popular prostitute was well beyond the pocket of the average patron—could always buy a poster and take it home. The most popular images went through multiple editions. Lautrec brought poster art to France, and today we have the pinup. Marilyn Monroe stands, legs splayed, skirt flying, over the same street vent on a million walls. Whether the *shunga* are erotic art or pornography is largely a question of context (context of production, consumption, and viewing); this is an issue we will come back to.

The developments in printing were not the only technological dimension of the Edo pornography industry. Both the *manga* and *shunga* specialized in representations of exotic and unlikely sexual practices involving remarkable acts of contortion and prowess between every imaginable combination of life forms. Variation and originality were important in an increasingly competitive market.

Sample Edo print.

The prints frequently involved any number of an assortment of popular sex aids of the day. The lists of items available for purchase by prostitutes and patrons at any of the local "knickknack" stores in the Edo brothel quarters (the largest quarter was the Yoshiwara district) would put even the most specialized of today's "adult accessory stores" to shame.[2]

The consumer audience of the Edo *manga* was predominantly male, and the humor is constructed around the phallocentric fantasy

Sample Edo print.

and desire of the male patron. More playfully put, these are phalla-cies. The same can be said of the woodblock prints, certain genres of popular fiction (which included various styles of illustrated narra-tives considered by some to be the precursors of the modern Japa-nese comic book), and even, perhaps I should say especially, the sex aids. Prints of lesbian or male homosexual sex or of prostitutes (male or female) masturbating were extremely popular. Woodblock depic-tions of lesbian sex almost always involved some form of phallus sub-stitute, most often a dildo strapped to the body of one of the lovers. Images of a woman masturbating with a dildo—covered in animal skin or intricate carvings—strapped to her foot, while acknowledg-ing the active sexual desire of the female, ultimately reinscribe the phallus into the autoerotic moment, and draw attention to the cul-tural insistence that even if it doesn't take two to tango it does take one plus a phallus.

This phallo-technology was an integral component of the com-modification and commercialization of sex in the Edo period. That there was a degree of cooperation and entrepreneurship involved between the suppliers of the sex aids and the woodblock print shops is not documented, but is a reasonable assumption given the highly commercialized environment of the brothel districts. The same association is easily documented today in the pornographic comic books, where page after page of advertisements for increasingly technologized sex aids blur into the image-texts of the surrounding stories, where the "goods" are contextualized into the sexual practice and fantasy represented there. While the complexity of technology and design has shifted dramatically, the basic concept and function of the manual sex aids on the market in Edo and contemporary Japan have altered very little. Pleasure remains intricately bound to penetration, whether it be anal or vaginal, heterosexual, homosexual, or masturbatory. Today, the silk kimono and wooden dildo of Edo are transformed into leather and steel, with a complimentary rechargeable battery included with each order over 10,000 yen.

It was the Meiji period and the reopening of contact with the West that witnessed the first major crackdown on pornography in Japan. It was the desire to meet with the perceived moral standards of the West that would finally result in the introduction and enforcement of strict censorship codes to regulate the publication of both images and written texts.[3] While there were laws restricting pornography during the Edo period, they had little impact on production and circulation. Throughout the Meiji and Taisho periods there was extensive experimentation with Western-style comic strips and cartoons, but there was a very low tolerance of any explicitly sexual references. It was during World War II that the next examples of what might be called pornographic comics appeared. These were produced by the Japanese as propaganda to be dropped on the Allied troops in the Pacific. They showed scenes of G.I. Joe's wife "having a good time" back home while he fought the good fight on the front, or suffering as a widow at the hands of some lusty, ugly character.[4] Comics dropped on Australian and New Zealand troops depicted their wives being raped by larger-than-life G.I.s stalking the cities on R&R. The Occupation forces did nothing to stifle the rash of political comic strips that flourished in the climate of change immediately after the war. However, the Allied Command was not tolerant of anything that was considered disruptive to social morale, and this included any image that was deemed pornographic. While comics for children and political satire were extremely popular throughout the 1950s, it was not until the late 1960s that the first "erotic" images would find their way back into the comic repertoire.[5]

By the late 1960s Japan was well on the path to reestablishing its industrial base, had declared a national policy of income doubling, and was experiencing a period of rapid urbanization of its population and the breakdown of the extended family household in the wake of the emerging *mai-hoomu boomu* (my home boom) of the new nuclear household. Increased surplus income led to increased consumption of domestic appliances. By 1960, 90 percent of households owned television sets.[6] It was the rapid advent of cheap home entertainment through television that saw the demise of two of the more popular forms of children's entertainment of the 1950s, the street storytellers (*kamishibai*) and book lenders (*kashibonya*). The storytellers showed hand-drawn picture cards or scrolls in portable box displays while they narrated the events depicted in the box screen. There are examples of similar techniques dating back into the Edo period. In the 1950s the *kamishibai* performer would purchase or rent his pictorial repertoire from one of a number of companies. Few of these street performers produced their own artwork in the large cities. According to one estimate, there were some 10,000 *kamishibai* storytellers in 1953.[7] Book lenders were particularly popular in the Osaka region in the 1950s. What comic books were on the market were still only in hardback and relatively expensive. The system of book lenders who rented out books and comics for a low price was one way of overcoming the cost for many comic readers. Both these occupations suffered greatly from the introduction of television.[8] It was clear by this time, however, that there was a potential market for less expensive comic books even in the age of television.

The first of a new wave of postwar comic books produced by the major publishing houses were aimed at a target audience of teenage males. The most successful of these was *Shonen Magazine,* which was launched by Kodansha in 1959 and had already reached weekly scales of one million by the mid-1960s. Other major publishing houses were quick to jump into the market, a market that has proven itself insatiable, with the more popular weeklies and monthlies for boys now regularly selling three to four million copies an issue.[9] Here again the rapid commercialization of a specific form of popular culture is linked to technological developments in the printing industry, as was the case with woodblock printing in the Edo period. Not only were innovations in the printing industry important to the rapid turnaround of large print runs of weekly comics, but the introduction of paper recycling also helped reduce industry overhead and allowed the publishers to help underwrite the financial risk of over-production in their market saturation strategy. The weekly visit of the paper trader remains a regular feature of urban life in Japan. When the housewives hear the familiar sound of his jingle blaring over the loudspeaker of the minivan, they carry their piles of newspapers and comic books out to the street to trade them for cheap recycled tis-sues or toilet paper. The paper trader then sells the newspapers and comics to the recycling industry.

After the dramatic success of the comics for boys, the industry was quick to recognize the next potential market. *Shojo Friend* and *Mar-garet* were the first two comic magazines for girls to achieve a com-parable level of success. Throughout the 1950s and early 1960s, the majority of the artists of the comics-for-girls were male.[10] The domi-nant formulaic structure of these early comics for young girls was a quest that carried the female protagonist along a trajectory of adven-ture or romance toward either the happy ending of a union/reunion with the object of longing or the tragic ending of a separation through the loss or death of that object. Tragic endings were in the minority. Narrative progression usually traced a process of recupera-tion or naturalization of a female protagonist from a location outside the normative structuration of gendered subjectivity—orphan re-unites with parent(s), ugly duckling turns into swan, tomboy swaps jeans for skirts, boy-hater finds true love. While at the level of story the heroine is positively transformed and/or achieves her goal, at the ideological level she is inserted into her proper place as the property or object of desire of the male.

By the late 1960s women began to break into the ranks of the comic artists. As teenage girls who had grown up as devotees of the *manga* began to seek a place for themselves within the industry, the narrative structures of the comics-for-girls became far more complex and moved beyond the normative boundaries perpetuated by the male predecessors of this new generation of female comic artists. The first major shift came with *Seventeen*'s serialization of Mizuno Hideko's "Fire" from 1969 to 1971. The hero is an American rock star by the name of Aaron. It was this story that depicted the first sexually explicit scenes in the postwar comic books. Aaron pursues a life of sex, music, drugs, and violence. His career as a rock star is ended when his hand is badly injured in a violent encounter with the Hell's

Angels. It has been argued that the story was in fact a quite conservative morality tale, with Aaron meeting an appropriately sad ending in a ward of a mental institution.[11] This reading, however, ignores the fact that for two years the readers of "Fire" anxiously awaited the next installment of Aaron's wild life. There are stories of queues of teenage girls waiting for the next issue to hit the shelves of the bookstores.

The serialization of "Fire" marked several significant shifts in the comics-for-girls. The most obvious shift was to a male protagonist. The object of longing of the female protagonists of the earlier comics was usually male, but the female remained the central focus of the narrative. What is more, the male characters were consistently above reproach, unlike Aaron. Mizuno created a male protagonist who was neither the boy next door nor a shining prince. Aaron was given all the physical characteristics of the princely role, but it was his non-conventional, rebellious behavior that Mizuno developed to attract and hold her readership. The anticipation of sexually explicit images and references was also clearly an important motivation for the teenage female readership.

The shift to male protagonists took a further turn with Ikeda Riyoko's "Rose of Versailles" (*Margaret,* 1972–74). In this work, heterosexual love was replaced by male homosexual love, complete with "bed scenes," as they came to be known in Japanese (*beedo sheenu*), depicting young homosexual couples. It is somewhat problematic to describe the "bed scene" in "Rose of Versailles" as homosexual. The protagonist, Oscar, is a girl who has been raised as a boy by her/his military family. Oscar eventually ends up becoming a member of Marie Antoinette's personal guard and falls in love with a nobleman called Von Ferson. In a complicated and playful scene of gender confusion, Oscar dresses her/himself up in splendid evening dress and spends a romantic evening in public with Von Ferson, never disclosing that the boy cross-dressed as a girl is really female after all. Unfortunately for Oscar, Von Ferson eventually declares his real love to be Marie Antoinette.

Another homosexual relationship develops between Oscar and André, the son of Oscar's childhood nursemaid. Oscar wins André's lifelong devotion and love when she/he saves his life. When Oscar finally reveals her/himself to be female the story takes still another turn. André's love for Oscar is based on his homoerotic desire for the

Beedo sheenu (bed scene).

person he perceives to be a beautiful young man. Oscar's declaration that she/he is female is presented as a narrative escape from the "dilemma" of the homosexual relationship—the normative transformation of homosexual love to heterosexual—the solution is a false one, however, for there is an even greater barrier to Oscar and André's future together, and that is class. The female Oscar cannot marry someone as lowly as André. The story ends with the death of both— André at the barricades and Oscar at the Bastille.

"Rose of Versailles" plays endlessly with gendered identity and the relationship between that identity and sexuality, disrupting the myth of biology as destiny. Gender is mobile, not fixed, in this story. In 1971 Moto Hagio published a short story titled "The November Gymnasium," which explored the love-hate relationship of two beautiful young boys, Eric and Thoma. The narrative tension is sustained at the level of the suppressed homoerotic desire of the two boys. They are like two sides of a coin; where Eric is strong willed and violent, Thoma is gentle and loving. It is eventually revealed that they are indeed twins. At the end of the story, Thoma dies. The only moment of physical contact Moto allows the two is a brief embrace. However, in 1974 Moto reworked the story of Eric into a much longer serialized comic titled "The Heart of Thoma." In the later work Moto developed the theme of homoerotic love more openly, as she continued the story of Eric and his relationship of love and rivalry with three other youths after the death of Thoma. Ikeda's story opened the way for a whole new genre of *bishonen* (beautiful young boy) comic stories of homosexual love. These stories of homosexual lovers pursuing one another across an exotic and fantastic landscape of mountains, forests, chalets, and palaces are a far cry from the comic books read by American teenage girls in the 1960s.

It was the *bishonen* comics that first broke the public taboo on the representation of sex in the *manga* in the late 1960s and the 1970s. This trend saw the emergence of the magazine *June,* which specializes in *bishonen* stories. In 1990 the target readership of *June* remains teenage girls, but there is little doubt that it now has an extensive crossover readership that includes a significant gay male following. For a period in the 1970s the comics-for-girls were a major testing ground for the censorship laws. In addition to the homoerotic *bishonen* stories, the popular and influential comic artists branched out into stories of lesbian love and increasingly explicit representations of heterosexual sex. What made it possible for the comics-for-girls to go as far as they did was the so-called bed scene. As long as the characters did not roll over or come out from under the covers completely, there was no technical breach of the law, which specifically concerned itself only with the display of pubic hair and penis. The bed sheets crept further and further off the body, and occasionally an artist would risk a standing embrace or full-body profile. Buttocks survived the scrutiny of the censors and became a permanent feature by the early 1970s.

The *bishonen* genre continues to be extremely popular among the comics-for-girls on the market today. Sales of popular biweekly and monthly titles remain in the millions. *June* continues to be one of the major selling comics-for-girls as the industry moves into the 1990s. Its sales figures are undoubtedly boosted by the number of gay and crossover readers. The *June* layout is now standardized and relatively predictable. Each month's edition includes a mixture of serialized comic stories, still shots from recent movies with either a homosexual story line or a popular androgynous actor (the July 1989 issue included stills of Rupert Everett in *Another Country,* Jeremy Irons in *Dead Ringers,* assorted stills of Hugh Grant, and Prince's *Batman* video), reproductions of advertisements featuring male bodies (the same issue included Hugh Grant for *L'Uomo,* a French anti-AIDS campaign poster, and a French advertisement for Nike running shoes showing a group of male marathon runners standing exhausted in the rain), reviews of new record and video releases, listings of popular back issues (under the title *"June,* the Discrimination-Free Comic Magazine"), advice columns (including letters from teenage girls about their own heterosexual problems and inquiring about homosexual love, and also letters from gay males seeking advice on issues from safe sex to new gay clubs), advertisements for gay and transvestite clubs, and the regular feature of a sealed comic story. The sealed story is billed as the "Secret Series" and presented as the "hottest" of the stories in each issue.

The "Secret Series" is sealed on the outer edge and bound into the spine of the comic so that it cannot be read unless one purchases the comic and tears it out. This is an innovative technique on the part of the publishers to deal with the problem known as *tachi-yomi* (literally standing and reading—browsing). Bookstores are always filled with browsers flipping through the pages of the new releases in the comic book section. The images in the "Secret Series" are actually only marginally "hotter" than those in the rest of the magazine. The "Secret Series" does tend to have more violent images than is true for the rest of *June,* and some scenes represent sexual contact in a more sinister or threatening way than the usually highly stylized romanticism of homosexual love that has become the magazine's standard fare. The July 1989 issue's "Secret Series" includes the threat of cannibalism and the suggestion of fellatio, but neither is shown explicitly, only hinted at. The most "shocking" image of this image/text is that of two men kissing mouth to mouth. In the *bishonen* comics androgynous figures appear wrapped in each other's arms, naked in bed, passionately making love, gazing longingly into each other's eyes, and kissing each other on the cheek. However, images of the young lovers mouth kissing are rare within the *bishonen* genre to the present. A scene of a man kissing another man's lips is treated as being far more sensuous than any bed scene. The scene in the July 1989 "Secret Series" in which the mouth kiss occurs is clearly demarcated from the rest of the surrounding text by a shift from the usual white background of the frame to a solid purple background. There is no doubt that this is the one scene in this particular text that earns it the "secret" classification. The visual and narrative tension that develops around the homosexual mouth kiss is not unique to the *manga*. In representations of homosexual love in the Edo *shunga* there was a similar tension. The mouth kiss was associated with the most intimate or intense moments of sexual contact. This was true of both homosexual and heterosexual love, but by the 1960s and 1970s the influence of Hollywood movies had more than likely diluted the erotic value of the heterosexual mouth kiss, which now abounds in the *manga*, television dramas, and cinema. However, the tension surrounding this scene remained intact in the case of homosexual lovers.

The representation of sexual scenes in the *bishonen manga* has continued to rely upon innuendo and anticipation rather than explicit representations of sexual contact that would fall into any of the established categories of pornographic image, such as the money shot and the meat shot.[12] These "shots" would be contrary to the spirit of the *bishonen* genre and the expectations of its readers. The androgynous (usually) male characters explore relationships of equality that are free of domination and exploitation. There is often an exaggeratedly demonic "bad guy" character who acts as a dramatic point of contrast for the gentleness and equality of the relationship of the homosexual lovers. It is not unusual for a narrative to follow a male character in his discovery of the possibilities of homosexual love as he gradually transfers his affections from a woman to a man.

The word *jiyu* (freedom) occurs again and again across the pages of the *bishonen* comics. The feature story of the July 1989 *June* is called "All My Life" and follows the gradual shift of the older male protagonist's love from his girlfriend to a beautiful young boy. The story abounds in metaphors of freedom created around the older man's dream of opening an animal park. His desire to let the animals roam free, outside of cages and beyond bars, is a barely disguised reference to his own desire to live openly and outside the constraints of dominant social values. His young lover jokes that this would lead

Mouth kiss scene from *June* magazine "Secret Series."

to chaos with all the different species living together, to which he re-plies, "Yes, but they'd all be free." The standard narrative progres-sions of the *bishonen* comics trace the trajectory of sexual fantasies that go beyond the normative boundaries of gender and sexuality. However idealized or romanticized these love stories may be, they offer a rare respite from the dominant cultural production of same-ness (where difference—male/female—exists only as the guarantee of the continued privileging of a phallocentric construct of norma-tive heterosexuality). The objective of the *bishonen* narratives is not the transformation or naturalization of difference but the valorization of the imagined potentialities of alternative differentiations.

By the 1980s the market for the *bishonen* comics had expanded far beyond the original readership of pubescent schoolgirls to include gays, heterosexual male university students, and young women, in particular young *okusan* (housewives—literally, "the person at the back of the house"). One Japanese commentator writing on the *bis-honen* comics has suggested that the reason for their popularity among teenage girls can be traced to the unwillingness of Japanese schoolgirls to deal with a growing awareness of their own sexuality.[13] In other words, the *bishonen* comics amount to a denial of sexuality among teenage girls. Such a reading of these *manga* would itself ap-pear to be a denial of the sexual awareness and curiosity of the mil-lions of teenage girls who devour these comics the minute they hit the stands. If anything is being denied or rejected by the readership of the *bishonen* comics, it is the stringent construction of gender and sexual practice in postwar Japanese society.

Some Japanese feminists have suggested that the more affluent Japanese society has become, the more entrenched, but subtle, have become the mechanisms of gender organization operating across the society, from the law to education and from advertising to domestic architecture. For a brief moment in the *bishonen* comics of the 1960s and 1970s, a new generation of female comic artists and their read-ership teased out the possibilities of new identifications and tested the boundaries of differentiation. It is probably not a coincidence that the readership of the *bishonen* is constituted of groups within Japa-nese society that could be described as transitional. The schoolgirl's passage through puberty is a heavily monitored and controlled jour-ney from girlhood to womanhood. Within that society there are two states of womanhood—married and unmarried. The former marks a

Young male lovers *(June)*.

successful transition and the latter a failed transition. In the 1990s, the teenage schoolgirl continues to be educated for her role as "good wife wise mother," despite a national rhetoric of equal opportunity.[14] A popular educational best-seller for mothers asserts that "having only one science textbook ... ignores the logical minds of boys and the daily-life orientation of girls."[15] It goes on to suggest that textbooks for girls include practical daily-life examples of "how the wash gets whiter when you use bleach and how milk curdles when you add orange juice."[16] The same male educator suggests that young girls should always wash their own underwear (which should always be white) by hand; in this way a girl can build "pride and awareness connected with her sex, such as the fact that her body has the capacity to create new life, that she must carefully preserve that function until the day of her marriage ... how clear it makes a girl's heart to always keep her underwear clean."[17]

The fact that a recent Japanese study showed that 45 percent of the mothers surveyed want their sons to go on to four-year universities while only 18 percent answered so for their daughters is an indication that the level of gender discrimination represented in the above quotations is not limited to male educators.[18] The education system, the family, the media, the entire fabric of society shroud the female pubescent body, and her body is marked with the traces of the encounter. The individuated body is inscribed as the motherbody, as an organ of the body politic. In this context, is it strange that schoolgirls are so attracted to a fantasy world of nonreproductive bodies, as remarkably non-Japanese as they are nongendered, moving across a backdrop of a nonspecific landscape that is nowhere or, more specifically, that is anywhere that is not Japan? These same schoolgirls often continue reading the *bishonen* comics long after they leave school and girlhood for marriage and womanhood. As an *okusan* the young woman occupies the motherbody space. She is defined as nurturer to both husband and children. Her sexuality is circumscribed by the boundaries of the motherbody. Little wonder she seeks out the imaginary space of the *bishonen* comics in moments of disengagement.

There are some who go so far as to suggest that Japanese men are never required to pass through that process of separation that has been described as the most traumatic transition for the male — separation from the motherbody. The *nauui* (from the English *now* meaning *trendy* or *vogue*) Japanese word for this imperfectly oedi-palized male condition is *mazaa-kon* — mother complex.[19] Through his school life the young Japanese male continues to receive the love (always emotional, sometimes physical) and nurturance of the mother. Only when he has left high school is he expected and en-couraged to develop a "healthy" interest in the other sex and marry. The male then enters into a relationship with the motherbody of the *okusan*. In Japan the majority of male high school graduates enter university, and the ensuing four years constitute a transitional stage in which a relationship of desire and dependence is transferred from the maternal motherbody to the matrimonial motherbody.

The popularity of *manga* reading groups that specialize in collect-ing and exchanging *bishonen* comics has to be placed in the context of this process of transition. The *bishonen* comics offer a young adult male readership a fantastical space for the exploration of sexual de-sire outside the closed circuit of the oedipal theater of the family but on the familiar territory of the homosocial formations of their youth. The extent to which the motherbody dominates the male experience of puberty arguably augments the structures of male bonding that are such an overt aspect of male teenage experience in Japan. When the space that might be the site for the exploration of sexual desire of bodies other than the motherbody is foreclosed, male bonding within the traditionally accepted homosocial formations becomes the dominant alternative structure for the formation of intimate rela-tions. These homosocial formations continue to structure the social relationships of the Japanese male throughout his life — the mythol-ogy of the "comrade samurai" transposed onto the contemporary workplace.

The images and stories of androgynous and homosexual lovers might be read as an alternative site of sexual fantasy, where no-body is the motherbody, a brief respite from "healthy" interests. It is sig-nificant that in interviews with fans of the *bishonen* genre, the major-ity of the male students insisted that they did not consider the char-acters in these *manga* to be homosexual, arguing that this was a different kind of love from either heterosexual or homosexual love. At about the same time the young male university student graduates to take up his position as a company man, he is also expected to trade in his *bishonen* comics for images of heterosexual pornography. That the initial attraction of the *bishonen* comics was rooted primarily in their homosocial, and not their homoerotic, dimension makes the transfer from these comics to heterosexual, sometimes homophobic, pornography a simple progression.

It would be wrong, however, to suggest that all male readers of the *bishonen* comics are not attracted by the homoerotic dimension of these image-texts. This genre of comic books is a rare example of the depiction of homosexuality in contemporary Japanese popular culture. *June* has acknowledged the significance of this gay readership with a gradual increase of information and visual coverage of gay culture both in Japan and overseas. The gay readership is in some sense the group with the least complicated relationship to these image-texts. In a cultural landscape that remains otherwise generally hostile to overt representation or expression of the homoerotic, these texts offer gay readers a rare site for the possibility of a direct and positive identification without denial or modification. Increasingly, through the late 1970s and the 1980s, *June* has played a role in the construction of a collective gay identity in a society where older traditions of homosexual and bisexual practice have been lost to a puritanism modeled on the most repressive dimensions of Western law and morality.[20] Comics and magazines for a specifically gay market have followed, but *June* continues to cater to a diverse readership.

Although the various categories of readers—female teenagers, adult women, male university students, gay men—are located within quite disparate relations of pleasure to these image-texts, it is not difficult to see how for each group these narratives offer a desirable fantasy space. The *bishonen* stories do not contain desire (male or female) within normative "phallacies" of gendered subjectivity, but allow the imagination to take flight beyond the territory of homophobic phallologocentrism. The stories themselves and the fluid, often unframed images that are so characteristic of the genre open up a fantasy landscape onto which each reader is free to map his or her own topography of pleasure. This contrasts starkly with the genre of adult (a euphemism for pornographic) comics that emerged in the 1970s.

It was this new range of adult comics that would finally end the suspense that had built up around the hide-and-seek graphics of the comics-for-girls. Full frontals, meat and money shots, quickly became the stock-in-trade of these "adult" comics. The 1970s saw the rapid development of what is now possibly the fastest-growing section of the publishing industry in Japan, the pornographic comic book. There are two streams of pornographic *manga:* "comics-for-men" and "ladies' comics" (the Japanized English *reedeezu* often appears in the title). A 1989 survey of comic book stores and industry catalogues showed in the vicinity of 180 pornographic titles for either men or women were available in the summer of that year. This includes only weeklies, bimonthlies, and monthlies, and not the extensive range of special issues, reprints, and collected works.

Exact figures for the volume of sales of the pornographic comics are extremely difficult to establish. From the information I have been able to gather from basic market surveys in 1988 and 1989, I estimate that up to ten million comic books sold per month in Japan could fall into the category of pornography. Many of the companies now publishing pornographic comic books are small and highly specialized. These smaller firms are sensitive to any form of inquiry regarding sales or profits at a time when both the relationship between the police and the industry and that between the antipornography movement and the industry have sensitized publishers to all forms of scrutiny. The larger publishers—such as Kodansha, which publishes the weekly *Morning Comic,* and Shogakkan, which publishes perhaps the biggest-selling pornographic comic for men, *Big Comic*—are not prepared to categorize these publications as pornographic, and are unwilling to offer a breakdown of figures that would indicate the percentage of total profits gained from pornographic titles. Both Kodansha and Shogakkan are major publishing houses, marketing everything from children's books to the Japanese classics. Kodansha is subsidized by the government. As one antipornography activist commented, "The Japanese government has subsidized the tobacco and liquor industries, so why not pornography?"[21] What can be said, despite the lack of firm statistics, is that the pornographic comics are not an isolated or insignificant phenomenon in Japan.

Unlike pornography in North America, these pornographic comics are sold openly on bookstands, in convenience stores, and from vending machines. Peak-hour commuter trains are a sea of comic books, and many of them are pornographic. No one pays any more attention to a man reading a pornographic comic book than they would to someone reading a newspaper. Having said this of the male commuter, I should clarify by saying that women usually would not read the comics-for-ladies on a train. They read these *manga* at home or in such interim spaces (locations where the public and private intersect) as the waiting rooms of doctors' or dentists' offices, beauty parlors, and coffee shops. The comics-for-ladies open onto a whole other area of complexity in relation to questions of internalization. It is not possible in the space available here to discuss adequately both the comics-for-men and the comics-for-ladies, so I will confine myself to the former. I have discussed the comics-for-ladies in detail elsewhere.[22] What is important is that it is considered normal for adult males—and, more recently, adult females also—to read heteroerotic pornography. I will come back to the normative function of pornography.

Having used the word *pornographic* more than a few times, I must now attempt to define how I am using the term. Why is it that I describe the comics-for-men as pornographic but not the images of naked male lovers in the beautiful young boy comics? The distinction I am drawing between the two is not based simply on content, nor is it based upon explicitness. The antipornography movement in North America has tended to resort to these two criteria as the basis for defining pornography. The writings of Andrea Dworkin, Catharine MacKinnon, and Susanne Kappeler are representative of this approach.[23] Content- and explicitness-centered definitions of pornography, in their attempt to pin down a moving target, end up obscuring the contextual complexity and fluidity of pornography as it shifts and changes across boundaries of time and space (geographic, cultural, historical). By reducing the identity of the pornographic to issues of content, this approach has extended the range of the pornographic from "snuff" movies to television commercials and the Miss America Pageant.[24] Antipornography advocates in North America have taken up a campaign strategy not unlike conservative fear or contagion campaigns mounted around such issues as drugs, AIDS, stranger abduction, and so on. The pornographic is now everywhere, inescapable, an ever-present evil. Artists and filmmakers who attempt to experiment with new representations of the female body are constantly looking over their shoulders for the feminist enforcers of antipornography "standards." [25]

For the purposes of my work on Japanese comic books I have taken up Beverley Brown's characterization of pornography. Brown rejects attempts to extend the definition of the pornographic to encompass every image of the female body found anywhere: "Merely characterizing the pornographic in terms of explicitness is about as useful and as accurate as characterizing capitalism as extreme misery." [26] Her own attempts to define pornography are less an exercise in pinning down the enemy than a recognition of the futility of trying. Brown moves beyond the simplistic restriction to questions of explicitness to explore the complexity and diversity of variables that distinguish pornography from nonpornography. She describes the difference between pornography and a medical photograph as pornography's "non-transparent features, the elements which constitute it as a distinctive representational genre—a certain rhetoric of the body, forms of narration, placing and wording of captions and titles, stylisations and postures, a repertoire of milieux and costume, lighting effects etc." [27] There is room within Brown's treatment of the generic forms of pornography for the consideration of contextual variation implicit in memories of a child peeping into sex education books in search of a forbidden image or the masturbatory possibilities of a medical textbook. A content-based analysis of images cannot accommodate the contextual dimension of the relationship between image and pleasure.

To return to the *manga* themselves—the comics-for-men and comics-for-ladies present themselves self-consciously to the reader as pornographic.

> Pornography essentially provides a stock of visual repertoires
> constructed out of elements of the everyday, using objects,
> including elements of the feminine, already placed within definite
> cultural practices. In re-placing these objects, in making them
> available or special, as objects around which sexual fantasy can
> operate without too much wit or effort, pornography
> simultaneously opens up the possibility of a reversal, of seeing
> objects return to their cultural niches with a certain afterglow.[28]

These comics transform the subjects, objects, and experience of the everyday—commuting, eating, golf clubs, tea ceremony utensils, baseball bats, food processors, secretaries, bosses, plumbers, and much more—into eroticized, fantasy objects. The distinction between the stories in these two genres and the stories in *June* is the primary objective of the transformation of the mundane. The stories in *June* are located in an idealized space outside of the "real." The goal is not the transformation of the mundane, but transportation be-

Replaced objects of the everyday.

yond the mundane to the fantastic. It is essential to the success of *June* that fantasy and the mundane never meet, that they remain two distinct worlds. The opposite is true of the pornographic comic genres.

I suggested earlier that this new genre of adult comics that came onto the market in the 1970s broke a tension or suspense that had

A graphics of nonrepresentation.

built up around the body graphics of the *bishonen* comics. While it is
true that the new pornographic comics abounded in full frontal images of male and female bodies and close-up money and meat shots,
they remained within the confines of the law forbidding the public
display of the penis or pubic hair. The artists of the adult comics
took the elusive and suggestive graphic techniques of the *bishonen*
(e.g., the entwined legs of lovers that make explicit what is hidden by
the carefully placed configuration of limbs) and developed an innovative and legal response to the letter of the law. To attempt to define
these comic images as pornographic on the grounds that they are explicit is problematic. This is a graphics of representation through
nonrepresentation. The pornographic meaning of a scene is never in
the image; it is produced in that interactive space between the image/
text and reader. An American printer who refused to allow an article
of mine that included scenes from a violent pornographic comic
story to go to print defended his decision by saying that he could not
print explicit representations of sexual organs. When it was explained that there are no organs present in the images, he responded

that this was even more perverse—and still refused to go to print. The question of which is more or less "perverse" aside, what these images foreground is a fundamental weakness in content- and explicitness-based definitions of pornography, which ignore the fact that meaning cannot be isolated in the image/text.

In the Edo-period woodblock prints the phallus was everywhere. In the pornographic comic books it is nowhere, and yet it is everywhere. It is always present in its absence. The current legal requirement that the penis be "removed" has led to graphic innovations that reinforce the Lacanian assertion that the phallus is not equal to the penis; it transcends the anatomical, signifying the power that is the privilege of the bearer of that organ. The frequent resort to phallus substitutes in the comics can be interpreted as a mechanism for the reinsertion of phallic order, a graphic counter to the threat of memory posed by the fragmented male body. This is one possible reading. Another interpretation would be that the proliferation of phallus substitutes is also an important graphic and diegetic mechanism for the transformative function of pornography. Objects of the mundane temporarily appropriated into the pornographic moment are re-ordered and re-placed in their everyday contexts. Simon Watney describes this process as one of "smuggling" forbidden or censored desires into the mundane.[29]

This traffic in eroticized objects is dependent upon the ability of the reader to play his or her role in the production of pornographic meaning. While a *manga* image may contain all of the components of a pornographic scene, it is often the reader who orders those components into a narrative whole. Taken out of the context of the preceding and following frames, a single comic frame can often be unintelligible. What facilitates the production of meaning is the reader's ability to synthesize the censored, incomplete frame within both the narrative (drawn and written) sequence of a specific story and the context of the entire repertoire of pornographic codes operating within the comic genre. For example:

nurse	therapeutic treatment (sadomasochism)
schoolgirl in uniform	object of pedophilic desire
stiletto heels	bar hostess
candle	torture device
conch shell	cunnilingus

Reading the repertoire of pornographic codes: candle as torture device.

fish mouth	desire of breast
train or plane in motion	penetration
Japanese white radish	phallus substitution or erection

These are just a few examples of an endless list of ever-changing visual codes familiar to the "informed" reader of the pornographic comics. The repertoire is constantly shifting in direct relation to fashion, movies, current best-selling novels, popular television events or programs, news, gossip, even the seasons and national festivals.

The pornographic *manga* involve a highly interactive reading process in which the reader scans the pages of the image/text (average speed per page of 3.2 seconds), contextualizing the components of each frame into a coherent whole—filling in the gaps as a child would paint by numbers, only in the *manga* the pornographic codes substitute for numbers. It is a small step from this interactive reading

Karaoke (sing-along) bar illustration from a comic book story.

process to the interactive software of the "erotic playing game" "Penguin in Bondage." The mass of "informed" users created by the *manga* are a ready-made market for this software. The interactive process is taken one step further when the reader/player is given a blank screen and a package of variable conditions and "items" with which to construct a personalized pornographic tale. In the "love hotels" (where rooms are rented by the hour) that rise in all their architectural absurdity across every Japanese cityscape—silhouettes of spacecraft, Egyptian pyramids, the *Queen Mary,* King Kong, and countless turreted castles—rooms are increasingly fitted with an assortment of technological gadgets that allow the occupants to program their own desired fantasy environment—sound effects, smells, temperature control, light, background images (sometimes projected life size on all four walls or a ceiling). Pornographic codes abound in this technological production of multiple contexts.

The codes also inform the images of naked women projected onto larger-than-life video screens in the *karaoke* (sing-along) bars that line the streets of each town's bar district. The background music of a popular song plays and a customer takes up a microphone to sing the vocal part—everyone can be a star—and behind him, on the screen, is projected a pornographic video that has only a vague connection to the content of the song but a very clear connection to the current repertoire of pornographic codes. The singer/player knows how to read the visual narrative played out on the screen just as he knows how to read meaning into the blank spaces of a *manga* frame. As he interacts with the coded images of the video, he inscribes himself into the narrative. He strokes the female body with a large feather, rubs a vibrating dildo over the empty airbrushed space between the legs of the woman's body projected onto the screen. He chooses his items from the available selection at the bar and the projected female goes on "reacting" until the song ends and the image fades out.

If marital sex is defined as a primarily reproductive function and the *okusan* occupies the motherbody space in relation to both her children and her spouse, then there is little room for sexual desire in the domestic sphere. It is in the bars and clubs that Japanese men will often act out their sexual desires. The theatricality of the *karaoke* bars is in keeping with the performative, almost ritualistic, acting out of desire between the hostesses and customers. Angela Carter captures the typical bar scene in her essay "Poor Butterfly":

> The girls even go so far as to feed their large infants food. "Open up!" they pipe, and in goes a heaped forkful of raw shellfish or smoked meat. Unaware of how grossly he has been babified, the customer masticates with satisfaction. . . . And a hostess can hardly call her breast her own for the duration of the hostilities.[30]

Here too the motherbody remains the primary object of desire, but in the environment of the bar world, for however brief a time, the men can act out fantasies of total indulgence at the hand of a mother/lover. This theater is really only an extension of the original family drama, where a business suited Oedipus gets to have his cake and eat it too—reclining naked women, decorated with a feast of raw fish, hand-feed their customers, and for dessert there is a bar down the street where flesh-colored ice cream comes in the shape of breasts, two to a plate. Desire and difference are organ-ized and hermetically sealed in a closed circuit of technoporn.

In the *bishonen* comics gender difference is displaced by gender mobility and androgyny. Normative or naturalizing narrative structures give way to narratives that follow a fantasy trajectory beyond the boundaries of dominant sexual identification and practice. It is not surprising that since the 1970s, while there have been unsuccessful attempts by conservative coalitions to have the *bishonen* comics removed from the shelves, sealed in plastic, and restricted to over-the-counter access, the sale of the pornographic comics has received almost no scrutiny from conservative interest groups. It is the *bishonen* and not the pornographic comics that are perceived as potentially disruptive of the fabric of society, opening up a fantasy space where there is a potential for the dis-organ-ization of the individuated body from the body politic. Unwittingly, the conservative campaigns point to the inherently normative and conservative character of dominant pornography. Feminists, gays, and lesbians warn against the tightening of censorship laws as a means of containing pornography in Japan, for they foresee their materials as far more likely objects of censorship than the images of sexual violence of the comics.

An antipornography campaign that emphasizes content and explicitness serves to reinforce the false sense of pornography as aberrant, positing the possibility of representations that are not fragmented and objectifying, and sexual practices that are "wholesome" and "good." The former is attractive but difficult to imagine and the latter is easily confused with the rhetoric of the Moral Majority et al. Other approaches, such as Brown's, allow us to map the common ground on which pornography intersects with the dominant constructions of gender identity and sexual practice. Once these sites of intersection (e.g., law, mass media, medicine, education, welfare, taxation) are identified, such strategies as Dworkin and MacKinnon's resort to legal controls for the elimination of pornography become less than convincing. Only a mad dog bites its own tail. Pornography and censorship share the same function of foreclosing potential sites of alternative identifications. They both order and organ-ize bodies.

The hide-and-seek graphics of the *beedo sheenu* of the comics-for-girls were the first attempt on the part of the comic artists to represent sexual contact while remaining within the limits of the censorship ban on pubic hair and penis. The artists of the comics-for-men took the graphic techniques devised in the comics-for-girls and went beyond the hide-and-seek approach of carefully arranged bed sheets and body parts to create an interactive reading process not unlike a jigsaw-puzzle effect. If the guide to the jigsaw puzzle is the photo on the box lid, the guide to the pornographic comic is the reader's familiarity with the current repertoire of codes. The comic reader has an ever-growing variety of titles to choose from. He can buy a comic for almost any fantasy or sexual preference he may have. The comics abound in advertisements for myriad sex aids and toys ("items") for his personal enjoyment, including life-size dolls with "vagina mouths" dressed in the "uniforms" of standard female stereotypes from the *manga,* vaginal molds with biographies and life-size photographs of the models, and electronic (waterproof!) "three-point" (anal, vaginal, and clitoral) stimulators. While many of these "items" are not unique to Japan's sex industry, what is unique is their ready accessibility through a medium as popular and pervasive as the comic books. In the bars the choice of pornographic videos is updated from week to week, and new "items" are added to the props for the interactive pornographic audio-video sing-alongs.

With the proliferation of choice, difference is disappeared. The choices mask the impossibility of deviance if that deviance threatens to dis-order or re-place dominant constructs of pleasure and power. The choices are always "informed" choices. Technology lends itself to speed, and the processes of recuperation accelerate in turn. A technique that began in a comic genre that at least attempted to tease out other potential identifications became the technographic starting point for a pornography that insinuates the reader into the graphics of the narratives of its image-texts, playing with (even depending upon) the reader's ability to make the "right" choices. Where the comics-for-girls are potential sites for the exploration of difference, the comics-for-men act as mechanisms of sameness. Japan's new age of technoporn literally captures the imagination and fantasy of the male consumer. The proliferation of choice adds to the complexity of the maze, but there are finally only a finite number of "correct" choices and then it's back to the beginning and round we go again and again. Technology provides the variation necessary to keep the players engaged—repetition without boredom—but the boundaries of the game remain constant. Technology achieves a captive audience. The player is no more likely to leap beyond the bounds of the game than the female character bouncing across the video screen of the latest YPG.

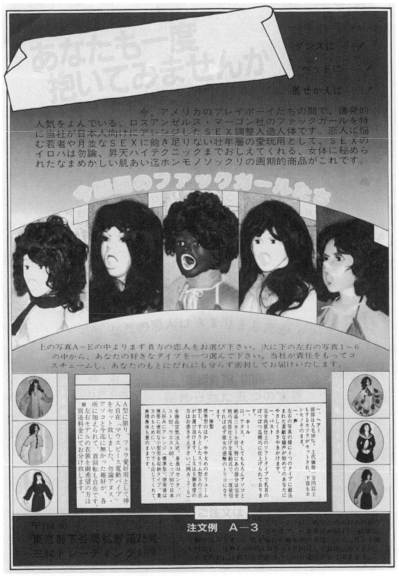

Comic book advertisement for sex aids.

NOTES

1. For more detail on Hokusai's *manga,* see A. H. Mayor, *Hokusai* (New York: Metropolitan Museum of Art, 1985).

2. The most detailed account in English of all aspects of prostitute life in the brothel district is the eyewitness account of J. E. De Becker, *The Nightless City* (Bremen: Max Nossler & Co., 1899).

3. See R. H. Mitchell, *Censorship in Imperial Japan* (Princeton, N.J.: Princeton University Press, 1983).

4. For an extensive bibliography of English sources on wartime propaganda, see L. D. Meo, *Japan's Radio War on Australia 1941–45* (Melbourne: Melbourne University Press, 1968), 290–93.

5. For overviews of the history of *manga* there are limited English sources. For some discussion see Tsurumi Shunsuke, *A Cultural History of Post-war Japan, 1945–1980* (London: Kegan Paul International, 1984). Also see F. Schodt, *Manga Manga* (Tokyo: Kodansha, 1983) and I. Buruma, *Behind the Mask: On Sexual Demons, Sacred Mothers, Transvestites, Gangsters, and Other Japanese Cultural Heroes* (New York: Pantheon, 1984).

6. Fuse Akiko, "The Japanese Family in Transition Part 2," *Japan Foundation Newsletter* (October 1984), 4.

7. Schodt, *Manga Manga,* 63.

8. Tsurumi outlines the relationship between the *kamishibai* and the comics; see *A Cultural History,* 41.

9. Details of the emergence of comics in the 1950s are drawn from Takatori Hide, "Shojo maga ni okeru ai to sei" (Love and Sex in Comics-for-Girls), *Dacapo,* 164 (1988), 6–11.

10. Ibid.

11. Ibid., 6.

12. Linda Williams discusses these sex-trade terms in detail in her excellent study of the genres of pornographic film, *Hard Core: Power, Pleasure, and the "Frenzy of the Visible"* (Berkeley: University of California Press, 1990).

13. Ibid., 21.

14. "Good wife wise mother" is the translation of a traditional Japanese term, *ryosai kembo.*

15. Quoted and translated in Mack Horton, "Reactionaries on the Shelf: Advice to Japanese Mothers by Gentlemen Amateurs," *Feminist International,* 2 (1980), 30.

16. Ibid.

17. Ibid.

18. *Survey of Attitudes to Education* (Prime Minister's Office) (Tokyo: Japanese Government Publications, 1982).

19. The term *mazaa-kon* was coined and popularized by the Japanese feminist Ueno Chizuko.

20. With the reopening of Japan in the late nineteenth century, Japanese lawmakers were extremely conscious of Western opinion as they framed the legal codes. It was of great concern to the Japanese that they not be considered inferior or lacking in "moral stature." To this end the Meiji administration sought guidelines from European and American laws relating to such issues as sexual morality.

21. Quotation from unpublished interview with Funabashi Kuniko of Media Watch Japan (summer 1989).

22. Comics-for-ladies are discussed extensively in my *Phallic Fantasies: Sexuality and Violence in Japanese Comic Books,* manuscript in preparation.

23. Andrea Dworkin, *Pornography: Men Possessing Women* (New York: Perigee, 1981); Catharine MacKinnon, *Feminism Unmodified: Discourses on Life and Law* (Cambridge, Mass.: Harvard University Press, 1987); Susanne Kappeler, *The Pornography of Representation* (Minneapolis: University of Minnesota Press, 1986).

24. For an example of the application of this approach, see S. Gubar and J. Hoff, *For Adult Users Only* (Indianapolis: Indiana University Press, 1989).

25. From 1984 to 1985 *Fuse* carried a series of articles dealing with the tension between a feminist antipornography position and the anticensorship position advocated here from the perspective of artists.

26. Beverley Brown, "A Feminist Interest in Pornography: Some Modest Proposals," *m/f*, 5–6 (1981).

27. Ibid., 7.

28. Ibid, 11.

29. Simon Watney, *Policing Desire: Pornography, AIDS, and the Media* (Minneapolis: University of Minnesota Press, 1987), 73.

30. Angela Carter, "Poor Butterfly," *Nothing Sacred* (London: Virago, 1982), 48.

Hybridity, the Rap Race, and Pedagogy for the 1990s

Houston A. Baker, Jr.

A functional change in a sign-system is a violent event.

—Gayatri Spivak

> *Yes*
> *Was the start of my last jam*
> *So here it is again, another def jam*
> *But since I gave you all a little something*
> *That we knew you lacked*
> *They still consider me a new jack.*

—Public Enemy, "Don't Believe the Hype"

I

Turntables in the park displace the machine in the garden. Postindustrial, hyperurban, black American sound puts asunder that which machines have joined together . . . and dances . . . to hip hop acoustics of Kool DJ Herc. "Excuse me, Sir, but we're about to do a thang . . . over in the park and, like how much would you charge us to plug into your electricity?" A B-Boy, camp site is thus established. And Herc goes to work . . . with two turntables and a truckload of pizzazz. He takes fetishized, commodified discs of sound and creates—through a trained ear and deft hands—a sound that virtually commands (like Queen Latifah) assembled listeners to dance.

It was the "monstrous" sound system of Kool DJ Herc which
dominated hip hop in its formative days. Herc came from Kingston,
Jamaica, in 1967, when the toasting or DJ style of his own country
was still fairly new. Giant speaker boxes were essential in the

competitive world of Jamaican sound systems . . . and Herc
murdered the Bronx opposition with his volume and shattering
frequency range.[1]

It was Herc who saw possibilities of mixing his own formulas
through remixing prerecorded sound. His enemy was a dully constructed, other-side-of-town discomania that made South and West
Bronx hip hoppers ill. Disco was not *dope* in the eyes, ears, and agile
bodies of black Bronx teenagers . . . and Queens and Brooklyn felt
the same.

There are gender-coded reasons for the refusal of disco. Disco's
club DJs were often gay, and the culture of Eurodisco was populously
gay. Hence, a rejection of disco carried more than judgments of exclusively musical taste. A certain homophobia can be inferred—even
a macho redaction. But it is also important to note the high-market-
place maneuvering that brought disco onto the pop scene with full
force.

The LeBaron Taylor move was to create a crossover movement in
which black R&B stations would be used as testing grounds for singles headed for largely white audiences.[2] Johnnie Taylor's 1975
"Disco Lady" was one of the first hits to be so marketed; two and a
half million singles sold. And the rest is history.

What was *displaced* by disco, ultimately, was R&B, a funky black
music as general "popular" entertainment. Also displaced (just
dissed) were a number of black, male, classical R&B artists. Hey, some
resentment of disco culture and a reassertion of black manhood
rights (rites)—no matter who populated discotheques—was a natural thing. And what the early hip hoppers saw was that the task for the
break between general "popular" and being "black by popular demand" had to be occupied. And as Albert Murray, that longtime
stomper of the blues who knows all about omni-Americans, put it: In
the *break* you have to be nimble, or not at all![3]

Queens, Brooklyn, and the Bronx decided to "B," to breakdance,
to hip hop to rhythms of a dismembered, sampled, and remixed
sound meant for energetic audiences—in parks, in school auditoriums, at high school dances, on the corner (if you had the power from
a light post . . . and a crowd). And Herc was there before Grandmas-
ter Flash and Afrika Bambaataa. And hip hop was doing it as in-group,
urban style, as music disseminated on cassette tapes . . . until Sylvia
Robinson realized its "popular" general possibilities and sugared it

up at Sugarhill Productions. Sylvia released "Rapper's Delight" (1979) with her own son on the cut making noises like "To the hip hop, hippedy hop / You don't stop." The release of "Rapper's Delight" began the recommercialization of B-ing. The stylistic credo and cryptography of hip hop were pared away to a reproducible sound called "rap." And "rap" was definitely a mass-market product after "Rapper's Delight" achieved a stunning commercial success. "B-style" came in from the cold. No longer was it—as crossover/commercial—"too black, too strong" for the popular charts. (But, of course, things have gotten stranger and *2 live* since then!)

II

So, rap is like a rich stock garnered from the sudden simmering of titanic B-boy/B-girl energies. Such energies were diffused over black cityscapes. They were open-ended in moves, shoes, hats, and sounds brought to any breaking competition. Jazzy Jay reports:

> We'd find these beats, these heavy percussive beats, that would drive the hip hop people on the dance floor to breakdance. A lot of times it would be a two-second spot, a drum beat, a drum break, and we'd mix that back and forth, extend it, make it 20 minutes long.
>
> If you weren't in the hip hop industry or around it, you wouldn't ever have heard a lot of these records.[4]

Twenty minutes of competitive sound meant holding the mike not only to "B," but also to set the beat—to beat out the competition with the "defness" of your style. So . . . it was always a *throwdown:* a self-tailored, self-tutored, and newly cued game stolen from the multinational marketplace. B-style competed always for (what else?) consumers. The more paying listeners or dancers you had for circulating cassettes or ear-shattering parties in the park, the more the quality of your sneakers improved. The idea was for youth to buy your sound.

Herc's black, Promethean appropriation of the two-turntable technology of disco and his conversion of discotech into a newly constructed blackurban form turned the tables on analysts and market surveyors alike. For competing disco DJs merely *blended* one disc into a successor in order to keep the energized robots of a commercial style (not unlike lambada) in perpetual motion on the dance floor. *To disco* became a verb, but one without verve to blackurban youth. What Herc, Flash, and their cohort did was to actualize the immanent possibilities of discotechnology. They turned two turntables into a sound system through the technical addition of a beat box, heavy amplification, headphones, and very, very fast hands.

Why listen—the early hip hop DJs asked—to an entire commercial disc if the disc contained only twenty (or two) seconds of worthwhile sound? Why not *work* that sound by having two copies of the same disc on separate turntables, moving the sound on the two tables in DJ-orchestrated patterns, creating thereby a worthwhile sound? The result was an indefinitely extendable, varied, reflexively signifying hip hop sonics—indeed, a deft sounding of postmodernism.

The techniques of rap were not simply ones of selective extension and modification. They also included massive archiving. Black sounds (African drums, bebop melodies, James Brown shouts, jazz improvs, Ellington riffs, blues innuendos, doo-wop croons, reggae words, calypso rhythms) were gathered into a reservoir of threads that DJs wove into intriguing tapestries of anxiety and influence. The word that comes to mind is *hybrid.*

III

Discotechnology was hybridized through the human hand and ear—the DJ turned wildman at the turntable. The conversion produced a rap DJ who became a postmodern, ritual priest of sound rather than a passive spectator in an isolated DJ booth making robots turn. A reverse cyborgism was clearly at work in the rap conversion. The high technology of advanced sound production was reclaimed by and for human ears and the human body's innovative abilities. A hybrid sound then erupted in seemingly dead urban acoustical spaces. (By *postmodern* I intend the nonauthoritative collaging or archiving of sound and styles that bespeaks a deconstructive hybridity. Linearity and progress yield to a dizzying synchronicity.)

The Bronx, Brooklyn, Queens—called by the Reagan/Bush era black "holes" of urban blight—became concentrated masses of a new style, a hybrid sonics hip-hoppingly full of that piss, sass, and technological vinegar that tropes Langston Hughes, saying: "*I'm still here!*" [5] This is a *black hole* shooting hip hop quasars and bum rushing sucker, political DJs.

IV

What time was it? Time to get busy from the mid-seventies into the wild-style popularizations of the eighties. From Parks to Priority Records—from random sampling to Run DMC. Fiercely competitive and hugely braggadocious in their energies, the quest of the emergent rap technologists was for the baddest toasts, boasts, and signifying possible. The form was male-dominant . . . though KRS One and the earliest male posses will tell you the "ladies" were *always* there. Answering back, dissing the ways of menfolk and kinfolk alike who tried to ease them into the postmodern dozens. Hey, Millie Jackson had done the voice-over with musical backdrop—had talked to wrongdoing menfolk (at length), before Run or Daryl had ever even figured out that some day they might segue into each other's voices talking 'bout some "dumb girl." Indeed!

Rap technology includes "scratching": rapidly moving the "wheels of steel" (i.e., turntables) back and forth with the disc cued, creating a deconstructed sound. There is "sampling": taking a portion (phrase, riff, percussive vamp, etc.) of a known or unknown record (or a video game squawk, a touch-tone telephone medley, verbal tag from Malcolm X or Martin Luther King) and combining it in the overall mix. (The "sample" was called a "cut" in the earliest days.) There's also "punch phrasing": to erupt into the sound of turntable 1 with a percussive sample from turntable 2 by def cuing.

But the most acrobatic of the technics is the verb and reverb of the human voice pushed straight out, or emulated by synthesizers, or emulating drums and falsettoes, rhyming, chiming sound that is a mnemonic for blackurbanity. The voice is individual talent holding the mike for as long as it can invoke and evoke a black tradition that is both prefabricated and in formation. "Yo, man, I hear Ellington, but you done put a new (w)rap on it!" For the rap to be defly *yours* and properly original, it has got to be *ours*—to sound like *us*.

The voice, some commentators have suggested, echoes African griots, black preachers, Apollo DJs, Birdland MCs, Muhammed Ali, black street-corner males' signifying, oratory of the Nation of Islam, and get-down ghetto slang. The voice becomes the thing in which, finally, raptechnology catches the consciousness of the young.

V

What time is it? The beginning of the decade to end a century. It is postindustrial, drum machine, synthesizer, sampling, remix, multi-track studio time. But it is also a time in which *the voice* and *the bodies* of rap and dance beat the rap of technologically induced (repro-duced) indolence, impotence, or (in) difference.

Why? Because sales figures are a mighty index. But also . . . the mo-tion of the ocean of dancers who fill vast, olympian spaces of audito-riums and stadiums transnationally when you are (*à la* Roxanne) "live on stage" is still a principal measure of rap success. Technology can create a rap disc, but only the voice dancing to wheels of steel and producing a hip hopping, responsive audience gives testimony to a full-filled *break*. You ain't busted a move, in other words, until the audience lets you know you're in the groove.

VI

What time is it? It's "hard core" and "message" and "stop the vio-lence" and "2 live" and "ladies first"—1990s—time. Microcomput-ers, drum machines, electric keyboards, synthesizers are all involved in the audio. And MTV and the grammarians of the proper Grammy Awards have had their hands forced.

Rap is a too-live category for the Grammies to ignore, and Fab Five Freddy and *Yo! MTV Raps* have twice-a-week billing these days. Jesse Jackson and Quincy Jones proclaim that "rap is here to stay." Quincy has even composed and orchestrated a cross-generational album (*Back on the Block*) on which he announces his postmodernity in the sonics of rap. Ice-T and Big Daddy Kane prop him up "on every lean-ing side."

But it is also time to "fight the power" as Public Enemy knows— the power of media control. In their classic rap "Don't Believe the Hype," PE indicates that prime-time media are afraid of rap's mes-sage, considering it both offensive and dangerous. In Philadelphia, one of the principal popular music stations confirms PE's assessment—WUSL ("Power 99") proudly advertises its "no-rap workday." Secretaries fill a sixty-second ad spot with kudos for the station's erasure of rap. Hence, FCC "public" space is contoured in Philly in ways that erase the energy of rap's postmodern soundings. "Work" (defined as tedious office labor) is, thus, publicly constructed as incompatible with "rap." Ethics and outputs of wage labor are held to be incommensurate with postmodern black expressive culture. Implicit in a "no-rap workday," of course, is an agon between industrial ("Fordist") strategies of typing pool (word-processing pool?) standardization and a radical hybridity of sound and morals. For rap's sonics are disruptive in themselves. They become even more cacophonous when they are augmented by the black voice's antiestablishment injunctions, libido urgings, and condemnations of coercive standardization. To "get the job done" or "paid in full" in the economies of rap is scarcely to sit for eight hours cultivating carpal tunnel syndrome. Nope. To get the job done with rap style is to "get busy," innovative, and outrageous with *fresh* sounds and defly nonstandard moves. One must be undisciplined, that is to say, to be "in effect."

Eric B and Rakim, Twin Hype, Silk Tymes Leather, Kingpin Redhead, De La Soul, Q-Tip, The DOC—the names in themselves read like a Toni Morrison catalogue of nonstandard cultural denomination. And such named rap ensembles and the forms they produce are scarcely local or parochial. Rap has become an international, metropolitan hybrid. From New Delhi to Ibadan it is busy interrupting the average workday.

VII

3rd Bass is a prime example of rap's hybrid crossovers. The duo is prismatically white, but defly blackurban in its stylings and "gas face" dismissals of too-melodic "black" artists such as MC Hammer.[6] ("Holy Moly!" as a notorious media character used to say: white boys, and one of them a graduate of Columbia, dissing a melanin-identified black boy for being not black or strong enough.) Which is to say that "we" are no longer in a Bronx or Brooklyn or Queens era but at the forefront of transnational postmodernism. The audience begins at eleven or twelve years of age and extends, at least, through post-B.A. accreditation. Rap is everywhere among adolescents, young adults, and entry-level professionals. It is a site of racial controversy, as in the anti-Semitism fiascos of Public Enemy.[7] It is a zone of gender problematics, ranging from charges against the form's rampant sexism (2 Live Crew is too flagrant here) through the throwdown energies of Queen Latifah and her ladies first, to the irony of squeaky-clean Good Girls. It is a domain of the improper, where copyright and "professional courtesy" are held in contempt. Rappers will take what is "yours" and turn it into a "parody" of you—and not even begin to pay you in full. An example is N.W.A.'s "It's not about a salary" line signifying on Boogie Down Productions, whom, so I am told, they can't abide. Rap is a place of direct, vocal, actional challenge to regnant authority: N.W.A., again, with " . . . Tha Police." Class is also a major determinant in the rap field. Its postmodernity is a lower-class, blackurban, sonic emergent speaking to (as PE has it) "a nation of millions."

VIII

Microcomputation, multitrack recording, video imaging, and the highly innovative vocalizations and choreography of blackurban youth have produced a postmodern form that is fiercely intertextual, open-ended, hybrid. It has not only rendered melody virtually anomalous for any theory of "new music," but also revised a current generation's expectations where "poetry" is concerned. Technology's effect on student expectations and pedagogical requirements in, say, "English literature classrooms" is tellingly captured by recent experiences that I have had and would like to share. To prepare myself for a talk I was to give at New York's Poetry Project symposium titled "Poetry for the Next Society" (1989), I decided to query my students in a course devoted to Afro-American women writers. "What," I asked, "will be the poetry for the next society?" To a man or woman, my students responded "rap" and "MTV."

We didn't stop to dissect their claims, nor did we attempt a poetics of the popular. Instead, we tried to extrapolate from what seemed two significant forms of the present era a description of their being-in-the-world. Terms that emerged included *public, performative, au-*

dible, theatrical, communal, intrasensory, postmodern, oral, memorable, and *intertextual.* What this list suggests is that my students believe the function of poetry belongs in our era to a telecommunal, popular space in which a global audience interacts with performative artists. A link between music and performance—specifically popular music and performance—seems determinative in their definition of the current and future function of poetry.

They are heirs to a history in which art, audience, entertainment, and instruction have assumed profoundly new meanings. The embodied catharsis of Dick Clark's bandstand or Don Cornelius's soul train would be virtually unrecognizable—or so one thinks—to Aristotle. Thus, Elvis, Chuck Berry, and the Shirelles foreshadow and historically overdetermine the Boss, Bobby Brown, and Kool Moe Dee as, let us say, *people's poets.*

My students' responses, however, are not nearly as natural or original as they may seem on first view. In fact, they have a familiar cast within a history of contestation and contradistinction governing the relationship between poetry and the state. The exclusion of poets from the republic by Plato is the primary Western site of this contest. (One envisions a no-poetry workday, as it were.) In Egypt it is Thoth and the King; in Afro-America it is the Preacher and the Bluesman. It would be oversacramental to speak of this contest as one between the letter and the spirit, and it would be too Freudian by half to speak of it as a struggle between the law and taboo. The simplest way to describe it is in terms of a tensional resonance between homogeneity and heterogeneity.

Plato argues the necessity of a homogeneous state designed to withstand the bluesiness of poets who are always intent on worrying such a line by signifying and troping irreverently on it and continually setting up conditionals. "What if this?" and "What if that?" To have a homogeneous line, Plato advocates that philosophers effectively eliminate poets.

If the state is the site of what linguists call the *constative,* then poetry is an alternative space of the *conditional.* If the state keeps itself in line, as Benedict Anderson suggests, through the linear, empty space of homogeneity, then poetry worries this space or line with heterogeneous performance.[8] If the state is a place of reading the lines correctly, then poetry is the site of audition, of embodied sounding on state wrongs such as N.W.A.'s " ... Tha Police," or PE's

"Black Steel in the Hour of Chaos." What, for example, happens to
the state line about the death of the black family and the voiceless
derogation of black youth when Run DMC explodes the state line
with the rap:

> *Kings from Queens*
> *From Queens Come Kings*
> *We're Raising Hell Like a Class When the Lunch Bell Rings!*
> *Kings will be Praised*
> *And Hell Will Be Raised*
> *Suckers try to phase us*
> *But We Won't be phased!*

In considering the contestation between homogeneity and heterogeneity, I am drawing on the work of scholars Homi Bhabha and
Peter Stallybrass,[9] who suggest that nationalist or postrevolutionary
discourse is always a discourse of the split subject. In order to construct the nation, it is necessary to preserve a homogeneity of remembrance (such as anthems, waving flags, and unifying slogans) in
conjunction with an amnesia of heterogeneity. If poetry, like rap, is
disruptive performance, or, in Homi Bhabha's formulation, an articulation of the melancholia of the people's wounding by and before
the emergence of the state line, then poetry can be defined, again
like rap, as an audible or sounding space of opposition.

Rap is the form of audition in our present era that utterly refuses to
sing anthems of, say, white male hegemony.

IX

A final autobiographical instance of rap-shifted student expectations
on the pedagogical front will conclude my sounding of postmodernism. I recently (February 1990) had the experience of crossing the
Atlantic by night, followed by a metropolitan ride from Heathrow Air-
port to North Westminster Community School in order to teach
Shakespeare's *Henry V* to a class of GCSE (General Certificate of Secondary Education) students. Never mind the circumstances occasion-
ing the trip—no, on second thought, the circumstances are popularly
important. A reporter for London's *The Mail on Sunday* had gotten
onto the fact that I advocated rap as an absolute prerequisite for any

teacher attempting to communicate with students between the ages of twelve and twenty-five.[10] So there I was in London, in a school with students representing sixty-seven nationalities and speaking twenty-two languages, in the Paddington/Marylebone area. "Once more into the breach dear friends / Once more into the breach / or let us close the wall up with our English dead" was the passage the students were supposed to have concentrated on, paying special attention to notions of "patriotism."

Introduced by the head of the English Department to a class doing everything but the postmodern boogie on desktops, I pulled up a chair, sat down, and calmly said: "I've come from the United States. I've been awake for thirty-six hours, and I have to listen to you so that I can answer questions from my teenage son about what you are listening to, what you are *into*. So, please, start by telling me your names." Even as they began to give me their names (with varying degrees of cooperative audibility), a black British young woman was lining up twelve rap cassette boxes on her desk immediately in front of me. (Hey, she knew I had *nothing* to teach her!)

To make an exciting pedagogical story brief, we took off—as a group. I showed them how Henry V was a rapper—a cold dissing, def con man, tougher than leather and smoother than ice, an artisan of words. His response to the French dauphin's gift of tennis balls was my first presentational text. And then ... "the breach." We did that in terms of a fence in the yard of a house that you have just purchased. A neighbor breaches it ... "How, George? How could your neighbor breach it?" George jerked up from that final nod that would have put him totally asleep and said, "What?" "Could your neighbor do anything to breach your fence, George?" "No, sir, I don't think so." "Come on George!" "Sir ... Oh, yeah, he could break it."

And then the anterior question about "breaches" and "fences" was arrived at by another student, and I leaped out of my chair in congratulation. "Sir, the first question is 'Why was the fence there in the first place?' " Right! What time was it?

X

It was time for Public Enemy's "Don't Believe the Hype." Because all of that Agincourt admonition and "breach" rhetoric (the whole hybrid, international class of London GCSE students knew) was a function of the English church being required to pay the king "in full," and the state treasury can get the duckets only if ancient (and spurious) boundary claims are made to send Henry V and the boys into somebody else's yard. "Patriotism," a show of hands by the class revealed, is a "hype" if it means dying for England. Bless his soul, though, there was *one* stout lad who held up his hand and said he would be ready to die for England. My black British young lady, who had put her tapes away, shouted across the room: "That's because you're English!"

Hybridity: a variety of sounds coming together to arouse interest in a classic work of Shakespearean creation.

The Mail on Sunday reporter told me as we left North Westminster that the English Department head had asked her to apologize to me in advance for the GCSE group because they would never listen to what I had to say and would split the room *as soon as the bell rang.* What the head had not factored into her apologetics was the technology I came bearing. I carried along my very own Panasonic cassette blaster as the postmodern analogue of both "the message" and the "rapper's delight" that Shakespeare himself would include in his plays were he writing today. At a site of postmodern, immigrant, sonic (twenty-two languages) hybridity produced by an internationally accessible technology, I gained pedagogical entrée by playing in the new and very, very sound game of rap. Like Jesse, I believe rap is here to stay. Other forms such as "house" and "hip-house" and "rap reggae" may spin off, but "rap" is now classical black sound. It is the "in effect" archive where postmodernism has been *dopely* sampled for the international 1990s.

NOTES

1. David Toop, *The Rap Attack* (Boston: South End, 1984), 78.

2. Nelson George, *The Death of Rhythm and Blues* (New York: Dutton, 1989), 149–58.

3. Albert Murray, *The Hero and the Blues* (Columbia: University of Missouri Press, 1973).

4. Quoted in John Leland and Steve Stein, "What It Is," *Village Voice*, 33 (January 19, 1988), 26. Leland and Stein's article is one moment in this special issue of the *Voice* devoted to hip hop.

5. Langston Hughes, "Still Here," *Selected Poems of Langston Hughes* (New York: Alfred Knopf, 1969), 123. "I've been scarred and battered. / My hopes the wind done scattered. / Snow has friz me, sun has baked me. / Looks like between 'em / They done

tried to make me / Stop laughin', stop lovin', stop livin'— / But I don't care! / *I'm Still here!"*

6. Playthell Benjamin, "Two Funky White Boys," *Village Voice,* 35 (January 9, 1990), 33–37.

7. Robert Christgau, "Jesus, Jews, and the Jackass Theory," *Village Voice,* 35 (January 16, 1990), 83–84, 86, 89.

8. Benedict Anderson, *Imagined Communities* (London: Verso, 1983).

9. Bhabha presented his brilliant insights on hybridity and heterogeneity in two lectures at the University of Pennsylvania on April 20–21, 1989. Stallybrass's essay on heterogeneity I read in manuscript, but it is destined for *Representations.* For the insight of Bhabha, one can read "The Other Question." In *Literature, Politics, and Theory,* ed. Francis Barker et al. (London: Methuen, 1986), 148–72. For Stallybrass, one might turn to his coedited monograph, *The Politics and Poetics of Transgression* (Ithaca, N. Y.: Cornell University Press, 1986).

10. Clarence Waldron, "Could Students Learn More If Taught with Rap Music?" *Jet,* 77 (January 29, 1990), 16–18. I had the honor of featuring prominently in this article. My picture even appeared in a gallery including Kurtis Blow, Kool Moe Dee, DJ Jazzy Jeff and the Fresh Prince, and Run DMC.

Watch Out, Dick Tracy!
Popular Video in the Wake of the
Exxon Valdez
DeeDee Halleck

Dick Tracy used advanced technology. He was the comic hero with the two-way radio wristwatch. He was also a cop, and cops can usually get the technology they need. But in 1948 I, a knobby-kneed eight-year-old girl, had a Dick Tracy watch, which made me the most technologically advanced of my family, not to mention on my block. No one in our neighborhood even had a TV set at that time. I got my two-way radio watch by sending in Kix (or was it Shredded Wheat?) box tops with a quarter and a self-addressed, stamped envelope. It was a classic case of military research benefiting the consumer. The cops got their equipment and we Kix eaters shared in the advancement of science. I was ecstatic. It didn't matter that it didn't work. Neither did the infinitely more frustrating battery-operated walkie-talkie my younger sister got in the late fifties. The Dick Tracy watch had no pretensions. It *was* pretend. I didn't expect it to work. It was adapted for the home market. It was the idea that counted.

From the cereal box to the TV set, the military is a part of everyday life in the United States. The merger between GE and RCA has only made more obvious the kinds of symbiosis that the military has had from the beginning with the major media corporations.[1] General Electric and RCA have been two of the biggest military contractors since World War II. NBC was a subsidiary of RCA and is now a part of the megamilitary corporation the merger created. Collusion of military and communication technology isn't new. The first large U.S. corporation was Western Union, which had its western expansion subsidized by Congress as a wartime expense for the Union during the Civil War. Native Americans fought this expansion by cutting the wires and on occasion felling the telegraph poles with axes. Resistance to the relentless advance of corporate communications seems just as futile today, if not more so. What recourse exists in a world ringed with satellites and stitched with microwave links?

Western Union told the Indians that it had "singing wires." [2] It didn't tell them that the wires could be used to call for troop reinforcements to suppress resistance. Information from transnational information corporations about *information* is obviously suspect. The views of IBM on artificial intelligence or of Kodak on the history of images are complicitous with the profit needs of their corporations, whether these views are expressed in a public relations handout or more subtly woven into a museum exhibit.[3] However, there are some notions about communication and technology that are so pervasive that they are accepted as axioms and are not so easily perceived. One such view of technology is that military research is the source of advanced consumer products. We are to feel grateful to the Joint Chiefs of Staff for everything from Tang (the space-age orange juice) to snowmobiles. I recall that Stanley Kubrick's film *Barry Lyndon* was heralded with the news that miraculous advances in lens technology made by the U.S. Army (chasing Vietnamese peasants in the dark, no doubt) enabled certain scenes to be shot in candlelight. In more recent years, news clips have shown how interstate bridge builders are able to use specially designed army helicopters (ideal for machine gunners hovering over banana plantations and coffee terraces).

It's not always the army. Sometimes it's the U.S. space program (a largely military exercise, however) that is our benefactor. The space/military complex has been credited with the development of the video camera and the digital watch. Technocrats like the trickle-down theory of technology to boost military/space budgets. But the image of technology that is developed by and for experts with happy side benefits for the masses ignores the pressures and demands of the mass market in the development of such items as the digital watch and the VCR.[4] The notion that military research is a source of consumer benefits not only helps to boost budgets for General Dynamics, it also serves to mystify both technology and the military and remove both from any connection to popular need except in the most remote, secondhand way. This trickle-down disinformation serves to increase the sense of alienation and powerlessness of media work-

ers, artists and activists alike. If communication technology is fortified within the military, it is further from our reach. This construct
obscures some of the contradictions present in technology designed
for popular consumption, and in particular those inherent in the
mass-marketed apparatuses of communication.

This is not to imply that a research based on seduction of consumers is more progressive or necessarily better than military research.
Tang is a miserable excuse for orange juice, whether it is concocted
in the labs at Livermore, AT&T, or General Foods. But it may be instructive to contemplate what the idea of military research as consumer benefactor does to the way we think about research and the
potential for shaping that work. In the fallout from Three Mile Island,
Bhopal, and Chernobyl there is a distrust and distaste for technology
in general. The profound alienation and impotence that most people
feel about technology has overshadowed any embryonic thoughts
we might have had about the liberatory potential of most machines.
But people are more willing to struggle against nuclear power proliferation than against the toxic effects of our communication system.
The widespread sense of technological impotence is increased by
maintaining the myth that the development of communication technology is inherently based in a military arena. It is clear that everyone
but the radical right and the corporations have been effectively intimidated. How can we challenge the media if RCA/GE is in charge?

While it is important to understand that telecommunications in
this country has been framed and circumscribed by military *and* consumer research, both are driven almost exclusively by instrumental
arrangements of the media/military corporations in their quest for
profits. There is perhaps no country in the world in which broadcasting and telecommunications in general have been so devoid of public service directive. The history of how this evolved has been elo-
quently and exhaustively detailed by Erik Barnouw in his classic
three volumes, *A History of Broadcasting in the United States.*[5] Barnouw was not optimistic about the possibility of regulatory reform of
this system:

> There are those who appear to think that our commercial system
> can be reoriented by codes and review commissions, but the
> history of radio and television gives little reason, if any, for faith in
> these devices. Against such influences, built-in economic drives
> have always reasserted themselves without difficulty. The need for a

supplementary system based on other motives is paramount and crucial. . . .

The conflicts of interest built into our broadcast media probably guarantee that present arrangements will continue for some time. The advertising industry will not readily give up its custodianship of our cultural life, which it has purchased with good money. Our clandestine warfare agency will not readily surrender its purse strings over international transmitters. Major military contractors will not readily give up their central position in our principal communication medium. Congressmen will not readily surrender their right to accept benefits from an industry over which they legislate.[6]

Since Barnouw wrote these lines in the 1960s, there have not been any power shifts to offset his dim view, and in fact the levels of greed and corruption have risen to unpredicted heights. However, developments within cable and consumer technology have made the situation more complex. With the introduction of home video and public access television, there has been a media *evolution,* if not a video *revolution.* Video technology is being used by a vast number of people in ways that have begun to challenge the passive consumption model that has dominated electronic communication ever since department stores first began to sponsor radio concerts to sell sofas and radios over the air (for listening to those same concerts in one's own home).

Bertolt Brecht's famous treatise on the emancipatory possibilities of radio drew its inspiration from the inherent "democratic" potential of transmission itself.[7] There is no reason electronic transmission—broadcasting as opposed to receiving—cannot be a popular-based activity. There is no law of the apparatus that decrees that it is only *reception* of radio signals that is widespread. *Broad*casting (including radio and television and cable distribution and satellite transmission) always contains the potential of multiple transmitters: multiple sites of transmissions and multiple messages to be transmitted.

One writer who clearly perceived a dialectic in technological development is Lewis Mumford. In his classic exploration, *Technics and Civilization,* he notes the contradiction between capitalist development of technologies and their inherent social tendency toward democratization: "If we wish to retain the benefits of the machine, we can no longer afford to deny its chief social implication: namely, basic communism." [8] Mumford believed that this basic premise could be the source of a redirection of science toward a utopian social sphere in which technology would be at the service of human needs. Mumford's optimistic exhortations were written before Auschwitz and the atomic bomb, but his is not the work of a naive dreamer. It was written after the Depression, and the basic formulation stems from a deep comprehension of the social depths to which a profit-driven system could plunge. There are passages that sound like descriptions of Reagan/Bush Amerika:

> Labor has lost both its bargaining power and its capacity to obtain subsistence: the existence of substitute industries sometimes postpones the individual but does not avert the collective day of reckoning. Lacking the power to buy the necessaries of life for themselves, the plight of the displaced workers reacts upon those who remain at work: presently the whole structure collapses, and even financiers and enterprisers and managers are sucked into the whirlpool that their own cupidity, short-sightedness and folly have created.[9]

Mumford's solution was a rational approach to technology:

> All this is a commonplace: but it rises, not as a result of some obscure uncontrollable law, like the existence of spots on the sun, but as the outcome of our failure to take advantage by adequate social provision of the new processes of mechanized production.[10]

Mumford's vision was typical of the social prescriptions that animated much of the New Deal. This ideology breathes through the films from that period—*The City, The River,* and *Valley Town,* in which an abiding faith in technology (i.e., architecture, urban planning, and hydroelectric control of river valleys—TVA) reaches a crescendo like the soaring music of Aaron Copland on the sound track.[11] The cynicism of the present era makes the hyperbolic rhetoric of those films as silly as a tractor-trailer full of Edsels on the deck of the *Exxon Valdez.* Progress indeed. TVA evolved into a shoddy nuke proliferator only now confronting the legacy of toxic radioactive wastes it has generated throughout the region. The glorified suburban utopias of that same period are now stifling ghettos of disrepair and desperation, and many of the urban expanses bulldozed for city renewals are weed-choked wastelands that have given up awaiting the scarce capital for housing projects and have become the site of tin warehouses: "free-enterprise zones" where sweatshop foremen can bypass health and safety regulations and pay under-minimum offshore wages.

Against the backdrop of this history, the following question confronts us today: Are there essential differences between electronic and mechanical apparatuses that change the obviously bleak prospects for popular use of technology? In other words, is it possible to have a populist vision of the processes of *electronic* production? Mumford's technocratic vision has ominous implications in our postmodern age. His recipe for utopia called for central planning, rational state control of resources and labor. In an age of computerization, electronic surveillance, and multinational culture industries, the specter of any sort of "central planning" is frightening. The use of computer data and video images to arrest and persecute the Chinese students of Tiananmen Square is as horrifying as the high-tech tanks currently being designed for optimum crowd control in Israel and South Africa.[12] Electronic technology has enabled the notion of central control to be functional to a degree never envisioned by Mumford, Bellamy, or Fritz Lang. Only Kafka had a gloomy enough vision to foresee accurately the grotesque forms we now encounter. However, Mumford's sense of the contradictions inherent in the mass market and the active desires of people revealed his tremendous faith in the creative needs of humanity (or "mankind," in his prefeminist language):

> Creative life, in all its manifestations, is necessarily a social product. ... To treat such activity as egoistic enjoyment or as property is merely to brand it as trivial: for the fact is that creative activity is finally the only important business of mankind, the chief justification and the most durable fruit of its sojourn on the planet. The essential task of all sound economic activity is to produce a state in which creation will be a common fact in all experience.[13]

This is something that Sony well knows. As Ben Keen has pointed out:

> It should be understood that Sony was formed as an engineers' company; as such, it was pervaded by a strong, idealistic belief in the beneficial nature of modern technology. This may have been initially a reaction against the wartime squandering of technical creativity on weapons of destruction, but the fact that even today these ideals still seem to be important to the company suggests that they run somewhat deeper. It appears that the founders of Sony really did (and do) believe that their technological efforts could

significantly enrich people's lives. . . . By choosing unique product goals that involve immense engineering challenges (and hence pleasures), Japanese management has continually redefined the meanings of technological progress in its chosen field.[14]

Today Mumford's radical technocratic vision of central state control presents a chilling model of power. The challenge is to develop Mumford's insights into emancipated uses of technology in a decentralized and genuinely democratic way. Sony's totally administered and engineered corporate utopia may be even more frightening, but there are many other features of the consumer video phenomenon that are worthy of our attention. In fact, it is evident that pockets of resistance have arisen that have the potential to evolve into more highly organized and autonomous centers of democratic communications.

Equipment manufacturers have every interest in selling ever-increasing numbers of machines, and, to guarantee a mass market, the machines have to be priced low enough for everyone either to have one or to feel that one is not entirely out of reach. The level of competition has ensured that television picture quality has developed at increasingly higher levels and at ever-lowering prices. However, the corporate program suppliers (i.e., commercial television and the movie companies) have an obvious interest in maintaining control over the programs that are consumed and distributed, and, more important, over whether and how they are duplicated. One of the reasons U.S. firms lost the initiative in the magnetic duplication field was that they were so bent on pushing the videodisk. Laser disks can be stamped out for mass-market consumption and are completely passive: they cannot record cost-effectively for home consumption. But these passive play-only machines never gained a foothold in the home market. People may be willing to buy one-way passive digital disks for listening (these disks have completely taken over the audio field—try buying a record these days), but there is a difference between audio and visual attention; one can listen and *do* other things. Passive video technology has never made it with consumers. RCA wasted $550 million finding that out.

The masters of video technology have been successful at marketing elements of active control even for hitherto passive viewing. Early Sony ads stressed the "time-shifting" possibilities: one could change the flow of television transmission to watch specific programs selectively, at one's convenience. One of the most attractive features of a VCR is the control it gives to viewing: still frame, fast forward, rewind, repeat—these are controls that allow a selective and critical viewing in an active mode. VCR users are unwilling to rest as passive viewers. The consuming public demands that the machines they buy produce *and reproduce* and have the capacity to time-shift images.

From the beginning of photography, equipment manufacturers have attempted to encourage a certain type of image production. In the early days of consumer photography, equipment was sold to the home market by promoting portraits of family members, family milestones, get-togethers, and remembrance of travel. This "domestication" of imagery was continuous with the domestic fate of women: safe in the home, the kitchen, and occasionally on vacation. For almost an entire century a certain (usually gender-specific) passivity has been promoted in the imagery of Kodak ads and especially in their instruction books,[15] where almost 100% of the *subjects* for the novice camera users are women.

In their advertising and hobby books, Eastman Kodak never suggested that Brownie owners take pictures of their workplaces, or that they record their rank-and-file strikes. Nor did Kodak promote the use of still photography to document water pollution or industrial waste (especially since they are on the top of the list of industrial polluters).

With the development of the "home movie" camera, the subjects were the same as those in still photography, only now they waved to the camera. Eight-millimeter movie cameras and super-8 rigs were marketed as live-action snapshots. With the introduction of consumer video, home camcorders, like still photography and home-movie cameras before them, are pushed as adjuncts of the bourgeois heterosexual family.[16] Ads for cameras feature baby's first step, Mom greeting the proud freckle-faced catcher of a string of brown trout, and the tearful hugs of high school graduations. In the TV ads families and friends use camcorders in traditional "home movie" genres.

However, there are several areas of difference between video camcorders and snapshot or even home movie cameras: the first difference is the *size of investment*. Video cameras are more expensive than Brownie cameras, and consumers are apt to want to get more use from them than a yearly glimpse of the kid's birthday.

A second difference is in the *site of display*. Photos fit neatly into special albums piled on coffee tables and in gilt frames on bureau tops. In these almost sacred locations for memory and ceremony, homemade photos (and commissioned photos such as wedding pictures) are clearly kept apart from other commercial, documentary, and industrial photography. That type of "professional" picture is in magazines, in books, and on billboards—in other words, in public spaces. Home movies and slide shows are family rituals for holidays and milestone events. The projector is taken out of the closet, the chairs moved around, the screen erected, and the room darkened. The viewing is an extended family ceremony. Homemade video, however, has its place not in a separate and ceremonial realm. It is shown on the family TV set, which, although it is located as the "hearth" of the modern home, is the central receiver of external "reality," the window on the world outside; in this respect, it can be defined as a public space.

A third difference between home video and snapshot photography and home movies is the *use of sound*. The fact that video has voice makes it a different medium from most consumer film rigs (whose sync sound provisions never really worked for consumer use). With the introduction of sound there is an increased demand on the subject. It is embarrassing just to stand around waving at the camera if there is a microphone on. One needs something to say. Consequently, home videos tend to have more content.

There is also a *narrowing of the gap* between "amateur" images and "professional" ones. With home movies, we always knew they weren't MGM. The size of the image, the clarity of registration, the skillful use of lighting all made "the movies" look very different from home movies. With video that differential is less evident. The new video cameras obviate the use of lights, and the resolution of detail is approaching that of some network programming. So too the use of a hand-held style, casually zooming in and out, has appeared in IBM commercials and on MTV. All this has brought a meeting of the images that puts in doubt any attempt to hierarchize "professional" video photography.

Finally, an important difference has been the *expansion of the video market beyond individuals to organizations and groups*. Video is purchased by businesses and groups in a way that never happened with still photography and home movies. Golf clubhouses,

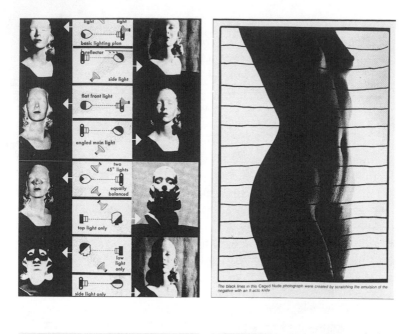

The black lines in this Caged Nude photograph were created by scratching the emulsion of the negative with an X-acto knife

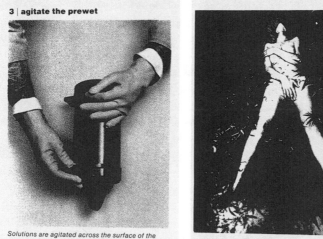

3 | agitate the prewet

Solutions are agitated across the surface of the paper by rolling or rotating the drum. Use a consistent pattern of agitation to prevent unwanted variations in prints.

The Sabattier effect, or solarization as it is commonly called, produces unique results but is hard to control. However, when done to film it enables you to make many identical prints.

Two-page spread from Diane Neumaier's "Teach Yourself Photography" (installation, 1987). Courtesy the artist.

Shutter Speed 1/500 Second	Shutter Speed 1/250 Second			
Bright or Hazy Sun on Light Sand or Snow	Bright or Hazy Sun (Distinct Shadows)	Cloudy Bright (No Shadows)	Heavy Overcast	Open Shade†
f/22	f/22*	f/11	f/8	f/8

*f/11 at 1/250 second for backlighted close-up subjects.
†Subject shaded from the sun but lighted by a large area of sky.

Fig. 118. "Lucille." A well lighted face

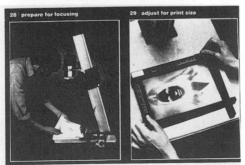

28 | prepare for focusing

29 | adjust for print size

day-care centers, ballet classes, and biology labs are potential camcorder users. Video has to be "user friendly" so as not to intimidate
the untutored; otherwise these groups would never feel comfortable
enough with the technology to purchase the equipment. As a result,
video handbooks break with the wife-and-kid-posing images that
adorn photo textbooks. Video sales pitches include shots of everyone from nuns to Cuna Indians shooting video. The stereotype of domestic use is broken. Consumer cameras show up at town council
meetings, school board tax hearings, rent strikes, block parties, and
Rotary Club meetings. Camcorders are becoming a fact of civic life
not easily dismissed. To be widely sold, video has had to be serious
(and fun), public (and private), and accessible across gender and racial boundaries. It has been marketed, in other words, as nothing less
than a popular tool.

The most far-reaching aspect of popular video use in the United
States has been the growth of the public access movement. Access to
channels and studio space and equipment is part of the cable fran-
chising process in cities and towns across the country. This movement has been underreported and misunderstood by both mainstream press and media critics. It is a grass-roots movement of
tremendous potential, although it varies a great deal in details from
city to city. In certain cities (Dallas, Somerville, Portland, Austin, Bur-
lington, Pocatello, and Atlanta are a few examples) a good mix of artists, media activists, community organizers, labor unions, and politicians has made public access an important, viable outlet for
community information, organization, and creativity.

In those cities where public access TV is thriving, there is also a
good audience for the access channels, or rather many audiences, for
when access works best, it is "narrowcasting"—providing programming for communities of interest, answering specific informational
needs. In Milwaukee, a group of deaf persons was able to produce a
weekly "signed" show so successfully that they inspired a group of
partially blind persons to get training and do a series themselves.
They received an oversized monitor through a grant to disabled persons and devised a system for using headphones for guests to signal
their camera cues, since the normal hand signals weren't readily perceived. Among the thousands of groups using public access are welfare rights groups, Latin America solidarity organizations, the United
Farm Workers Union, and local affiliates of Amnesty International.[17]

I have been involved with cable access since 1976, when I was part of a group called Image Union, formed to cover the Democratic National Convention. At that time we cablecast for five hours a night for five nights on Manhattan cable access. In the early days of video, when there were few portapaks, experimental work was done for the most part in collectives. These groups, such as Video Freex, TV TV, and Raindance, shared equipment and space and were the first groups to garner grants from the public arts councils as collectives. Although they made tapes for several years, few of the early video artists were actively involved in producing for television. Video art tended to be shown in closed-circuit situations, in galleries, museums, and media centers. A few public television specials highlighted this work (or it was shown in series, such VTR [*Video Television Review*] for WNET TV, curated by Russell Conner). These were ghettoized as arts magazines—sort of a "Believe It or Not" version of what was going on in the "art world." Their format did not present a challenge to the conventional forms of broadcast television.

However, by the late 1970s there were several alternative series running regularly on public access in New York: *Communications Update* (which I helped produce along with Liza Bear, Willoughby Sharp, Vickie Gholson, and Michael McClard), Colab's *Potato Wolf*, Glen O'Brien's *TV Party* ("A TV show that's a COCKTAIL PARTY and also a POLITICAL PARTY!"), and Jaime Davidovitch's *Artist TV*. These series were not designed for museum screenings, but were developed to be shown (often "live") as regular television programming. The audiences were small, but loyal. The most radical effect of these often wildly outrageous programs was to add a bit of leavening to the often weighty seriousness of public access talk shows.

In the initial years of access, most of the participants' energy and commitment went into guaranteeing space on the channels, rather than developing functional formats. All too often, producers for public access would mimic the commercial networks in pathetic attempts to reproduce the seamless effects their slick (and expensive) production values allow. In the late 1970s, public access video artists took up the challenge of creating cheap and easy television that mocked the artifices of the corporate TV sets. Flaunting procedural gaffs and deliberately skewing the rapid pacing of network timing, these programs provided a deconstruction of television that was at times quite literal. On *Mock Turtle Soup*, Image Union's coverage of the 1976 election night, the cartoonlike set began to fall apart. Predicting the moment when the cardboard painted clock on the wall would fall became more suspenseful than the election results from Minnesota.

In 1981, I was one of the founders of the access series *Paper Tiger Television*. These programs have been developed not only as programming on Manhattan Cable (and several other systems around the country) but as a model series for creative low-budget use of studio, small-format cameras, and local resources. The Paper Tiger Collective has now produced almost two hundred programs of media criticism, from "Herb Schiller Reads the *New York Times*" to "Donna Haraway on the *National Geographic.*" The ultimate value of this series lies less in the impact of any of the individual programs than in the overall effort to create television formats that are site specific to public access. As such, they have been useful examples of a kind of alternative television that makes virtue out of necessary schlock.

Because of this, Paper Tiger drew a number of enthusiasts from around the country, and we were able to make contact with other progressive access users, many of whom expressed the desire to exchange programming. It was out of these discussions that we were able to form the Deep Dish Satellite Network, a collaborative organization of access activists and producers, to share our programming via the commercial satellites. After the success of the first two series—the first of ten one-hour programs and the second of twenty hours, each fed weekly up to a satellite—the network has thrived, and there are plans for continuous weekly programming in the future. The programs are picked up by public access stations across the country and shown "live" or rebroadcast on local channels.

Most of the programs have been magazine-type shows, each tackling a specific social issue. For example, one program is called *Home Sweet Homefront.* Produced by Louis Messaih, it combines footage on the struggles for housing from many different communities, from Philadelphia, New York's Lower East Side, and Minneapolis, among others. The community video footage is ironically framed with Mumfordesque clips from housing films from the New Deal. The program neatly juxtaposes homeless activists with the liberal rhetoric of a bygone era. The show paints a vivid picture of a major crisis in locally specific terms. In direct contrast to the decontextualized and atomized way these issues are portrayed in the nightly network news, the local struggles are recontextualized in this program and given an additional historical frame of reference. Other Deep Dish shows have focused on the farm foreclosure crisis, pesticides, women's issues, and racism. Each combines local clips to form a larger national picture of the issue.

The shows have been popular on local channels, especially with overworked and underappreciated access volunteers who see the series as a valorization of the work they do in their communities. Since the series in essence is a "showcase" of good access programming, they often repeat the programs at a level of frequency not initially predicted by the network founders. For example, Austin Public Access did a "Deep Dish Month," after the regular weekly series was over. They programmed the first and second series together *nightly.* After that they did a "Deep Dish Weekend" during which they played Deep Dish back-to-back for thirty hours straight. Many cities across the country have likewise maximized the Deep Dish showings, often featuring articles about the series in their local program guides. Thus the series is a grass-roots phenomenon, in terms of its content, its distribution, and its promotion. Although local in origin, perhaps the most important effect is to valorize not only the work of local access producers, but the local struggles they document. Often these groups are isolated and alienated from their local communities. Deep Dish uses the technology to create communities of interest that prove to the video producers and the organizing groups that their work is part of a larger movement. Letters of support to Deep Dish have one phrase that is most often repeated: "Now we know we are not alone."

Deep Dish has also received letters from home satellite owners, a potential audience that now numbers more than four million. The majority of dish owners are in isolated rural areas without any other source of television signals. This individual satellite audience has been fully appreciated by Christian broadcasters, who use them for fund-raising and for proselytizing to other viewers. The right wing in this country has proved effective in the creation, through media technology, of an audience and a community that transcend geographic boundaries with technology.[18] The right's early use of direct mail and computer lists was only tardily replicated by environmentalist and antimilitarist groups. However, in recent years we have seen the successful development of Peacenet, a progressive computer network.

Peacenet provides electronic mail and computer data bases in such fields as environmental research, media analysis, Latin American refugee assistance, and antinuclear organizing. Many individuals and groups have come to rely on the circuits of data and exchange thereby provided. This network will be an important resource for any future networking possibilities in the video community.

In spring of 1989, I received a notice through Peacenet that a solidarity group in California was organizing a presentation in connection with the tenth anniversary of the Nicaraguan revolution: an hour interview with Daniel Ortega. The San Francisco group had set up down-links across the country (similar to the reception locations set up for boxing championships) in auditoriums in ten large cities. The message would come "live" to solidarity groups gathered for the celebration. After receiving the Peacenet memorandum via my computer, I contacted the San Francisco group and sent them the Deep Dish address list: more than four hundred stations that had picked up and played our last series. The group mailed out notices to this select list. This mailing, combined with the organization's own mailing list of solidarity groups and individuals, created a formidable network for the program. It was transmitted from Managua on July 22, 1989, and received by several hundred access stations across the country. The combination of computer networking, solidarity work, and video transmission formed a wide network that combined narrowcasting and broad information dissemination. This sort of consortium of organizations and expertise can be tapped in the future.

Why hasn't the left done this sort of thing before? Satellite networking has been commonplace among right-wing Christians for years. Although they have been able to con thousands of dollars by playing on the desperation of the faithful, most of the pledge dollars go into buying cars and split-level homes for the preachers. The actual dollars spent on the satellites and the technology are minimal. This sort of network is cheap when weighed against the price of mailing and duplication costs of more obsolete methods of program distribution. But the left has not made use of this opportunity. To this day, it is still uncommon for progressive groups to use technology in this way.

In the process of raising money for the Deep Dish series, I have had to address the question of why the left in the United States has not made use of potentially powerful tools for organizing and distribution of alternative media. Although in recent years there has been increasing willingness to critique mainstream media (the Institute for Media Analysis and Fairness and Accuracy in Reporting [FAIR] are two organizations dedicated to this purpose), there has been relatively little activity in the realm of creating alternatives to the official media. Issue after issue has been covered by individual films and videos, but there has been a reluctance to tackle broader distribution schemes.

The most active and savvy organizers in the realm of media are the public access activists throughout the country, but they are usually embroiled in local struggles and have little energy and few resources for extending their cause. However, there are opportunities out there. Deep Dish TV has been working with several other groups to initiate discussions about creating an authentic alternative network: a twenty-four-hour transponder[19] that will be a source for progressive programs and news. It is an uphill struggle. The resistance is not technological, but more ideological and financial. Struggling progressive organizations often resist spending money on equipment. Many groups are anxious about long-term commitments in lease agreements for even a photocopying machine. Only recently have public interest organizations been willing to computerize their mailing lists. Within the public interest community there is reluctance to use scarce funds for technology when more direct human needs are pressing. It seems more humane, more real, more pure to organize a rally or to create a refugee camp than to create a TV network. It is easier to get funds for a film about a coal strike than it is to fund a film about the lies the media are telling about the coal company. It is easier to organize a speaking tour than it is to organize the circulation of a television series. Unfortunately, the right in this country doesn't have these inhibitions.

Meanwhile, on main street, popular video has arrived and it is growing lustily. Video camcorders combined with community organizations have begun to cause ripples in the sands of the TV wasteland. The creative use of technology that Mumford dreamed of is alive in hundreds of small studios, in trailer parks, in community-controlled mobile TV vans, and in high school rec rooms. It's called public access. The rank and file are seizing the time and channels.

> One of the most interesting uses of video is as self defense against the police. For years African-Americans and Latinos have been victimized by excessive police force. Every year several hundred

young men die in police custody or in street struggles with undercover cops. Camcorder video has enabled communities to document these incidents. For years police have videotaped demonstrations and community organizations. But as mass sales of video recorders have increased, harassed communities have taken to watching the police. "It's the democratization of surveillance," said Larry Sapadin, Director of AIVF (The Association of Independent Video and Filmmakers).[20]

Watch out, Dick Tracy! We've got you covered.

NOTES

1. Jack O'Dell, chairman of the Pacifica Board, recently suggested that we should expand the term *military-industrial complex* to read more accurately *military-industrial-media* complex; "Anti-Communism in the U.S.: History and Consequences," conference held at Harvard University (1988).

2. Fritz Lang's first film in Hollywood was *Western Union,* in which Randolph Scott portrayed the cowboy who helped subdue the progress-resistant Indians.

3. Herb Schiller's *Culture, Inc.*(New York: Oxford University Press, 1989), documents how corporations have usurped public space for their messages.

4. The way in which much of consumer product history is attributed to military research is grist for the mills of the conspiracy theorists among us. Or, as Michael Parenti put it: "Why is it that everyone is willing to accept the notion that if you're talking about price fixing or contra funding we can recognize a conspiracy, but as soon as we mention the media in terms of ruling class needs we're all of a sudden conspiracy theorists?" Media Panel, "Anti-Communism in the U.S." conference at Harvard (November, 1988).

5. Erik Barnouw, *A History of Broadcasting in the United States* (New York: Oxford University Press, 1966).

6. Ibid., vol. 3, 339, 343.

7. Bertolt Brecht, "Radio as an Apparatus of Communication," *Brecht on Theatre,* ed. and trans. John Willett (New York: Hill & Wang, 1932), 51.

8. Lewis Mumford, *Technics and Civilization* (New York: Harcourt Brace, 1934), 406.

9. A recent *Nation* article describes how Iowa farm wives, caught in the spiral of loss and foreclosure, are jobbing auto part manufacture in the evenings on their farms.

10. Mumford, *Technics and Civilization,* 402.

11. Willard Van Dyke and Ralph Steiner, *The City* (1939); Pare Lorentz, *The River,* camera, Willard Van Dyke (1937); Willard Van Dyke and Irving Lerner, *Valley Town* (1940).

12. Maria Elena Hurtado and Judith Perera, with Jung-nam Chi and Mary Zajer, "The Science of Suppression," *South Magazine* (November 1988), 70: "A South African riot control vehicle, the Nongkai, was hailed by Pretoria's law and order minister, Adriaan Vlok, as a 'new concept in protection'—in this case protection of apartheid rule from unrest in South Africa's black townships. Equipped with a front-end loader, the Nongkai can remove obstacles like barricades and burning cars. It can also cut wire and even winch itself out of trouble. . . . Various methods of capture have been tested, in-

cluding launchers that fire nets over the crowd. Tear gas, intended ostensibly to disperse crowds, is often used in confined spaces and in dangerous concentrations. . . . The name Nongkai means peacemaker."

13. Mumford, *Technics and Civilization,* 410.

14. Ben Keen, " 'Play It Again, Sony': The Double Life of Home Video Technology," *Science as Culture* 1 (1987), 6–42.

15. Diane Neumaier, "Teach Yourself Photography: 50 Years of Hobby Manuals." Paper presented at the Society for Photographic Education Conference, San Diego (1987).

16. Sheila Pinkel, unpublished research on Eastman Kodak.

17. An overview of public access in one state is available from the Alternative Media Information Center, 121 Fulton Street, 5th Floor, New York, NY 10038. *The Participant Report on Cable Channels and Related Video Resources* is a county-by-county survey of who is using public access and what kinds of programs are being produced throughout New York State.

18. See the detailed look at the religious right in Sara Diamond, *Spiritual Warfare* (Boston: South End, 1989).

19. A transponder is the face or channel of a satellite that receives and retransmits a signal. Often a satellite will have twenty-four transponders. Although most transponders are already leased to commercial or right-wing religious television networks, there are many that are available for occasional rental, and a few that might be available for long-term lease for several years, twenty-four hours a day.

20. Constance L. Hop, "Sharper Focus in Videotaping: On the Police," *New York Times* (August 15, 1988), 14.

Just the Facts, Ma'am:
An Autobiography
Processed World Collective

I first saw a copy of Processed World . . . *about three years ago but had heard rumors about it along the disgruntled-clerical-worker grapevine for some years before that. In fact what I'd heard was that there was some anarchist magazine out in California somewhere that would tell you how temp workers could rapidly and secretly bring large corporations crashing to their knees via sabotage. So when the first issue I saw did not provide instructions for this project I was sorry, but liked it anyway.*

—A. R., New York (response to *PW* reader survey, spring 1989)

Processed World magazine was founded in 1981 by a small group of dissidents, mostly in their twenties, who were then working in San Francisco's financial district. The magazine's creators found themselves using their only marketable skill after years of university education: "handling information." In spite of being employed in offices as "temps," few really thought of themselves as "office workers." More common was the hopeful assertion that they were photographers, writers, artists, dancers, historians, or philosophers. But day after day, thousands of such aspiring creative types found themselves cramming into public transit en route to the ever-expanding Abusement Park of the financial district. Thus, from the start, the project's express purpose was twofold: to serve as a contact point and forum for malcontent office workers (and wage workers in general), and to provide a creative outlet for people whose talents were blocked by what they were obliged to do for money.

In the late 1970s, a number of radicals around San Francisco and New York who had ridden out the decline of the social opposition

231

with their brains unscrambled, principles more or less intact, and rage intensified found themselves drawn to the punk/New Wave milieu. Incoherent and often crude as it was, it looked like the only game in town—the only place where such fundamentals of the ruling ideology as Work, Family, Country, Obedience, and Niceness were being challenged with real *panache* and real venom. Some of these people formed bands or organized shows. Still others worked in graphic media, producing posters, fanzines, and comic books, or revived street theater and other kinds of political performance. Among these people, numbering at most a few thousand around the country, images, ideas, jokes, slogans, and techniques circulated like amphetamines in the cultural bloodstream.

Prior to the founding of *Processed World,* several participants had already shared in such activities. Lucius Cabins, in fact, first encountered *PW* cofounder Maxine Holz and early collaborator Louis Michaelson in a Bay Area agitprop group called the Union of Concerned Commies. The UCC began in late 1978 as a left-libertarian intervention into the antinuke movement, then at the height of its strength and militancy. The group tore apart within a few months, but well before the breakup the spirit that had animated the UCC was directed toward finding a new home. Before *PW* was even thought of, Cabins and Holz produced a leaflet for National Secretaries Day in April 1980 called "Innervoice #1" (under the name "Nasty Secretaries Liberation Front"). This leaflet foreshadowed the *PW* style. One side was a mock invoice listing the prices paid by an average office worker for her unhappy life. On the other side was a short analytical essay called "Rebellion Behind the Typewriter." It referred pointedly to the collective power of information handlers to subvert the circulation of capital.

A year later, in April 1981, the first *Processed World* hit the streets, Cabins and Holz having been joined by Michaelson and Christopher Winks (and Steve Stallone as printer) in producing that first issue. Finding themselves amid the bulging supply rooms of the modern office, *Processed World*'s friends began collecting resources for the magazine; the first two issues were printed on paper "donated" by San Francisco's major banks. A short while later, Gidget Digit and a half dozen others, mostly already friends of the founders, joined the newborn project. The cover art for *PW* 2 was drawn by a woman (Anne) who, with her co-worker, wrote the first wildly enthusiastic letter received by the magazine, and helped realize *PW*'s role as forum. Another new contact, Helen Highwater, frustrated with her efforts to write for the proto-union "Working Women's" newspaper *Downtown Women's News,* became an avid participant when she discovered *PW*'s hawkers on a busy downtown street while she was temping. Other participants came in the same way.

Processed World's founders saw the importance of communication among people in similar predicaments. Without a heightened sense of community no amount of rhetoric, agitation, or sabotage could begin to change conditions. Every Friday, writers and editors would head out to the streets to hawk magazines, asking a dollar donation rather than "selling" so as to avoid restrictions on street merchants and to remain protected by the First Amendment's freedom-of-speech provisions. Collective members would don papier-mâché costumes: VDT heads/masks labeled "IBM — Intensely Boring Machines" and "Data Slave"; an enormous detergent box with familiar red-and-yellow sides that read "Bound, Gagged, & TIED to useless work, day in, day out, for the rest of your life?" These attracted immediate if often puzzled attention from passersby. Sellers pranced around on busy financial district streets yelling *"Processed World:* the magazine with a bad attitude!" or "Are you doing the processing, or being processed?" or "If you hate your job, then you'll love this magazine!" (In fact, many of *PW*'s slogans were spontaneously composed on the streets.) In this way, *PW* managed to develop and maintain a fairly close rapport with its office work force readership. Many fascinating dialogues took place during these Friday lunch hours, and a feedback loop was established whereby readers, writers, and editors would discuss articles in person, right on the street.

More regular social events were developed in pursuit of a new dissident community. Biweekly gatherings at Specs bar in North Beach began in February 1982; these grew to attract upward of forty people every other Wednesday night, until they died out two and a half years later, in late 1984.

While a community of sorts was established, at least in fits and starts, it did not last. *PW* never had a specific goal in creating the community, beyond hoping for a movement to arise independent of the magazine. Many people came to the magazine and its gatherings looking for "answers," for some kind of organizational structure, or at least for an idea about what to do about their next working day.

But, while such ideas were plentiful, it was also felt that *PW* lacked a coherent "program"; it had no clear-cut plan of action in which to incorporate people, let alone an actual organizational presence in offices. Those who came looking for such things invariably went away disappointed.

For actual production *Processed World* depended on collating parties, a modern-day urban version of the barn raising. A week or two prior to the finish of printing, a leaflet would go out to the entire Bay Area mailing list, inviting people to come and help collate the forthcoming issue. Many people, isolated at their jobs, would eagerly come to the collating parties, seeing them as opportunities to contribute more than money to the *PW* effort, and also as their chance to meet and interact with the creators. From noon to as late as midnight (sometimes for two days), fifty to a hundred people would pass through and take a shift sitting before a tall stack of pages. Amid clouds of marijuana smoke, bottles of beer, and a rich potluck buffet, each person passed the folded pages along, adding his or her two or three sheets, until completed copies were boxed at the end of the line, ready for the bindery. Collating parties lasted through issue 18; by then, every joke about assembly lines and free labor had been thoroughly exhausted. *PW* finally went "upscale" and started paying to be printed and bound by a web-press company.

Single-minded dedication to one activity produces boredom. As an antidote to the paper and print world, *PW* began to explore other pursuits, sometimes as a way of advertising, sometimes as research, and always for fun.

On Saturday, April 14, 1984, *Processed World* organized a tour of Silicon Valley, hosted by Dennis Hayes, *PW*'s then-newfound SilVal writer. Piling into an old blue bus, some twenty-five malcontents visited a variety of strange locales in the fabled valley, by way of the U. S. Air Force's Blue Cube satellite control center, the Rolm Corporation's campuslike offices, and Benny Bufano's giant missilelike Madonna ("Our Lady of the Missiles"), to the squeaky-clean Fashion Island shopping mall. The tour got plenty of attention from nervous security guards and even received a weird write-up in *Infosystems* magazine, continuing *PW*'s already considerable—if erratic—media coverage.

Just a month later, the End of the World's Fair, a radical cultural festival aimed at providing alternatives to the New Age pieties of the mainstream peace and ecology movements, was held in San Francisco's Dolores Park. The organizing for the event was done out of the *PW* offices, though the committee behind the fair was independent of the collective. The *Processed World* Collective created two floats ("Terminals with Ears" and "Prisoners of Daily Life"), and, with many props, made a memorable appearance in the fair's parade and costume show. Annual picnics at various Bay Area parks provided another place for local people to meet informally. Sporadic benefits and poetry readings have also been staged.

The Playground

The world of the temp worker—one of arbitrary authority, transient workplace communities, and a cornucopia of technological invention—was the main object of the magazine's focus. Early articles examined the social world of the temp, that world's economic future, and the technology that prevails there. Science writer Tom Athanasiou provided insightful articles on the new information age (issue 1), technology's role in restructuring the world economy (issue 8), artificial intelligence (issue 13), and data encryption and privacy issues (issue 16). Other writers (Chris Winks and Brad Rose, to name but two) have examined subjects as diverse as video display terminals and their hazards (an ongoing topic), modern architecture and its functions, and toxic exposure. Dennis Hayes has contributed many informative and revealing articles on the hazards of the silicon industry (issues 10–23).

Although the technology deployed in the workplace defines certain limits—and certain freedoms—we maintain that it is the personal experience of the worker that is of paramount importance in understanding the high-tech workplace. From the first issue, *Processed World* has featured the "Tale of Toil," highlighting stories told by messengers, a secretary in a veterinarian's office, phone company workers, temp workers, a janitor, and a prostitute turned secretary. These worker narratives are hard to categorize: they are nonfiction in one sense, but are certainly not "objective," or the result of documentary research. However, they are among the readers' favorite features, and provide (at times) amazing insights into daily working lives.

Different, if perhaps more profound, understandings may be gleaned from the world of "fiction." From the first issue, with "San Francisco 1987—Would You Believe It?" which imagined an insurrection in the office district, the "unreal" has been given as much weight as the "real." Stories range from the erotic ("Kelly Girl Plays Postmistress"), to the horrifying ("Debth"—from *death* and de*bt*) to the surreal ("Thursday Morning" and "Naked Agenda"), to the sublimely hopeful (Michael Blumlein's "Softcore"). As a forum for fiction writers, *PW* has illuminated certain possibilities through the explorations of self and society.

Poetry has long graced the pages of *PW,* again ranging across a wide spectrum—Adam Cornford's fierce "Psalm of the Anger" and Tom Clark's "Pressures of the Assembly Line," which details the last minutes of a worker in a glass factory, contrast with Cornford's "Gringo Boy Poets" and its pointed comment on "language poets." From the prosaic to the esoteric, we have seen (among other things) hamburgers, shit, unemployment, and assassination. From a certain vantage point surrealism and capitalist realism merge.

PSALM OF THE ANGER

I   Because outside the ambulances howl at the dogs
II   Because the typist is forced to eat her own fingers
III   Because I come wrapped in cellophane and stamped with a blue number
IV   Because brain-damage leaves a little trail of wildflowers
V   Because we speak to each other only through a wire grille and our time is up
VI   Because even the forests are made to tell lies
VII   I want to crawl into the street soaked with burning oil
VIII   I want to smash clocks in my teeth and dig graves with my fingernails
IX   I want to spit out the pin of a grenade like a plumstone
X   I want to splatter the maps in the boardroom with bloody continents
XI   I want my screaming to dissolve cartilage
XII   I want my children's bodies to grow thick black fur

Adam Cornford, *Processed World,* issue 2

Reviews of poets, such as Antler, or of books (Langdon Winner's

The Whale and the Reactor, Earl Shorris's *Scenes from Corporate Life*), plays (Vaclav Havel, or the Artist and Audience Responsive Theater in San Francisco), and movies (*9 to 5* and *WarGames*) broaden the vantage point of both the readers and the producers of the magazine. The emphasis throughout these articles was on work and its subversion, on modern culture and what it tells us about our relation to labor.

Analysis of workers' responses to work—whether pondering sabotage, unionizing, or informal organizing—occupies many pages. From the cannery workers of Watsonville (issues 15–19), to the Blue Shield strike of 1980–81, to the Social Service Employees Union in San Francisco (an example of a nonbureaucratic union), to a lengthy analysis of an attempt to form a union in an art supply company, *PW* has a unique take on labor issues. The tension between a vigorous critique of traditional organizing goals and techniques and a recognition of the need for collective action has never been resolved. We are speaking here not simply of *how* workers are organized, but about their *goals*. For us the contradiction lies in favoring workplace organizing on the one hand, while on the other hand advocating the *abolition of work*. This paradox certainly is not new, but the abolition of work has never seemed so feasible (or so necessary).

Sabotage, which still excites media attention more than six decades after the IWW became infamous for its advocacy of worker resistance, has been a recurrent theme: Gidget Digit's "Sabotage, The Ultimate Video Game" (issue 5) and "CLODO speaks" (issue 10) analyze sabotage in the United States and in France. Workers who are isolated in their jobs, or trapped in an atomized culture, often find sabotage to be an effective form of opposition. There's nothing quite like it to bring back a bit of self-respect, especially among the most oppressed, or those bent on radically reconfiguring the power relations in "new tech" jobs. Sabotage is such a simple thing, widely practiced but rarely advocated, that it shocks people, both bosses and their erstwhile opponents. "Working to rule," absenteeism, poorly performed work, computer theft, and data destruction are all common occurrences, but "respectable" people seem only to decry such worker resistance, and never celebrate it, let alone the more conscious forms of revolt.

The creative and playful spirit that inspires the best of popular culture provides not only a shield for resisting the brutalizations of everyday life, but also a sword: the imagination that allows the reinvention of our world. In its emphasis on "culture" (i.e., by not being solely concerned with analysis and polemics), *PW* continues a rich tradition dating to the past century. The radical newspapers and magazines of the last century regularly printed poetry and fiction.

Unfortunately, the tendency to elevate "serious" (nonfiction) writing over the "frivolous" pursuits of fiction and poetry has been a dominant one. This parallels the ideology of the "professional" in the office—the unemotional, serious work of business is to be executed humorlessly. A polite smile of amusement is permissible; a guffaw is not. A polite frown *may* be allowed, but raised voices, unorthodox descriptions, and certain "tones of voice" are not.

Against this bland conformity of Serious People Doing Serious Things, we propose the whole range of emotion: exultation to despair, anger and whining, love and laughter. We receive a steady flow of letters that complain about the whining tone of some articles or other letters. Sure, we'd like our critical analyses to strike their mark clearly without irritating some readers, but our culture has increasingly frowned on expressions of discontent or "negativity," so any serious critique will invariably sound "whiny" to some. One of our principal aims, after all, has been to make people feel *good* about hating their jobs.

As the opening editorial in the first *Processed World* proclaimed: "Rebellion can be fun, and humor subversive." Every issue of the magazine has dedicated at least a third of its space to graphics, usually satirical. As part of its goal to be fun here and now, and to be an outlet for frustrated creative abilities, *PW* has given lots of room to graphic artists, collagists, cartoonists, and punsters.

What humor communicates is not simply the punch line or the meaning behind the joke, but also the pleasure of laughter itself. Aside from the sheer fun of it, the magazine's humor provides a more accessible, less direct way to express the attitudes and ideas put forth in the more "serious" articles. Humor has always been used to give vent to feelings and fantasies that are socially unacceptable or offensive, because jokes are less compromising than direct statements. The jokes themselves may be offensive, but ambiguous ("Does she really mean it, or is she just kidding?"). People who won't or can't

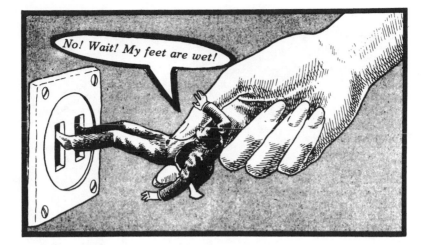

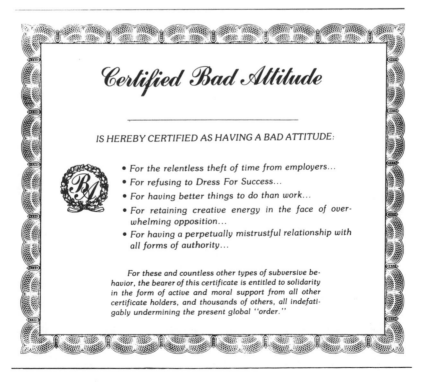

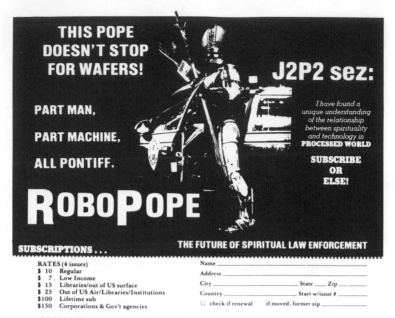

resort to open confrontation find an outlet in humor. Besides, many people don't form their critical perceptions of the world and themselves via rational, cognitive processes. The *directed* ambiguity of political humor can give people room to react and respond on other levels—attitudes, feelings, instincts.

At a time when North American radicals are hanging their heads, dizzied by the speed of negative developments, there isn't much in which to take pleasure. Or, conversely, everything is so painfully ridiculous that it inspires sardonic despair. But political humor provides an antidote to either kind of hopelessness, because it exudes a disrespect for What Is that implies people can change it. Sharing humor also reinforces the immediate subjective pleasure of life, which can occasionally be the basis for bigger, more serious collective endeavors, because it solidifies a sense of community among participants.

Nevertheless, humor plays an ambiguous role at work, providing a means both to reinforce and to undermine the authority of managers and routines. It can reinforce authority when it serves the purpose of

laughing off real problems instead of dealing with them (the ever-present office jokes about stale and toxic air). Such cracks about occupational health parallel the common jokes about carcinogens in food. The typical reaction is "Oh, doesn't everything cause cancer?" Hopelessness is so pervasive that most of us choose not to think about it, and, when confronted, we dismiss it with a cynical half-joke. It's pretty obvious who benefits from defeatist humor of that sort. Moreover, a lot of job-related humor is racist or sexist, homophobic or xenophobic, and therefore divisive.

Yet the workplace is also a natural laboratory for turning humor around and reclaiming its subversive spirit. *Processed World* developed a humorous discourse based on the imagery and language of the business world itself. Dozens of images were gleaned from the business and computer press (*Business Week, Fortune, Modern Office Procedures, Today's Office, Food Processing News,* and others) and then revealingly altered, or "detourned," to use the situationist

term. Sometimes these images and slogans are used in collage, but more often they have their overt message inverted or diverted by small additions or subtractions. A subjectively truthful caption changes the sense of a conformist image, or a bland corporate catchphrase is turned inside out by a bizarre or sinister graphic. Alternately, a graphic may be run as is—the fevered mind of the overtime ad writer may reveal more than was intended.

Processed World also uses humor because it serves to distance its own project from the deathly self-importance of the dogmatic leftists and their boring, oppressive ideas of "socialism." Seeking to encourage utopian thinking, to instill and legitimate aspirations for a world governed by pleasure and desire, *Processed World* cultivates its sense of humor at every opportunity.

Over the years, the focus has expanded beyond the office and its

Do people write advertisements about endangered species protection for multinational oil companies, who are among the worst polluters on the planet?

Chevron

PEOPLE DO.

graphic by Lucius Cabins

technologies. As its producers migrated out of clerical office work, the magazine began addressing a broader range of subjects, albeit without losing contact with its roots. Children (issue 14), food (issue 15), medicine (issue 20), militarism (issue 21), the environment (issue 22), and sex (issues 7 and 18) are some of the topics addressed in special issues that went considerably beyond the office while preserving a focus on work and its discontents.

As this is being written in autumn 1989, *Processed World* is pondering various futures. We have provided a good time for lots of readers and a social forum for participants, and we have shown that it is possible to do a good job *without* being paid for it. We have given thought to a more active role in the "green" movement as one possible way of becoming less "isolated." The nonprofit that owns *PW,* the Bay Area Center for Art and Technology (BACAT), is contemplating a leap into the video void—a current fantasy involves some form of interactive video robot on an expedition across America. The long flirtation between poetry and *PW* has resulted in a series of chapbooks currently being published by BACAT.

Another issue of *PW* is in the works—the magazine has never had a better guarantee than that. But the 1990s are upon us, and there is no need to keep repeating old actions. Our recent readers' survey made it clear that there is a place for a zine such as *PW,* and that we have, to some extent, broken out of the "leftist ghetto"—a fair number of our readers do not define themselves as radicals (Marxist, anarchist, or whatever). This in itself is a hopeful sign—if these ideas are *not* in everyone's head, they are at least surviving the grim dollar days of the 1980s.

Perhaps the best note to end on is taken from our collective editorial in issue 15 (Winter 1985-86):

> The magazine has gone in a different direction than the one its founders intended. *PW* was to be a meeting point for dissatisfied and rebellious workers in the "new" technical and service sectors, a place where they could vent their frustrations and share their dreams. So far, so good. But we wanted to go beyond frustration-venting and dream-sharing to help develop strategies for organized resistance at work. We wanted the rebellion to become *practical.*
>
> In 1980–81, this didn't look so far-fetched. Revolt was in the air. . . . But as the Right got a firmer grip on the mass media and as the recession hit, terrorizing millions of workers into submission, the revolt largely faded away. Today, an atmosphere of anxious subservience, thinly veiled in born-again patriotism and consumption-mania, pervades daily life.
>
> With office work in particular, the problem goes even deeper. *PW* has always distinguished its "take" on workplace organizing from more traditional approaches by pointing out that most work in the modern office is *at best* useless in terms of real human needs, and at worst (as with real-estate, banking, and nuclear and military contracting) actively destructive. Rebel office workers, sensing this, *don't identify with their work.* They generally change jobs often and work as little as possible. Their revolt takes the form of on-the-job *dis*organizing—absenteeism, disinformation, sabotage. They seldom view as worthwhile either the risk or the effort involved in creating a workers' self-defense organization. Moreover, rightly or wrongly, they believe that most workers, who identify more with their jobs, also identify with management. As a result, the rebels tend to be as alienated from their co-workers as they are from the boss. Perhaps this is why *PW*'s extensive discussions of autonomous office-worker organizing seem to fall largely on deaf ears—while its frequent references to sabotage have made it notorious. . . .

... Any real mass upsurge seems far away. In that case, isn't *PW* in danger of marketing the image of a non-existent revolt to be passively consumed by its reader-contributors? Perhaps. But we think that even in the absence of real revolt, *PW* is helping to create the cultural preconditions for it. Again and again, readers tell us: "I thought I was the only person who felt this way. Now I know I'm not alone." ...

... *PW* has always maintained that, beyond a culture of resistance and some organized self-defense against corporate and governmental power, we need a complete reinvention of the social world. ... Finally, it comes down to this. Through *PW*, we try to assert lucid imagination against Rambo-style reactionary fantasy, true diversity against careerist "individualism," free solidarity against authoritarian fake community, nameless wildness against well-organized death. This helps us to survive a bleak time. We hope it does the same for you. Together, perhaps, we can achieve a lot more. Write us.

The Processed World Collective
1829 Sutter St., #1829
San Francisco, CA 94104 USA

Understanding Mega-Events: If We Are the World, Then How Do We Change It?

Reebee Garofalo

Bob Geldof should be remembered in history for suggesting that a lot can be done if we tap this power source.

—Bill Graham, 1985

I would like it to be a movement, but it is not going to be so.

—Bob Geldof, 1985

In one of those moments that cried out for some grand social gesture, Joan Baez opened the Live Aid concert with the words: "Good morning, you children of the eighties. This is your Woodstock and it's long overdue." While there was undoubtedly a historical connection between the two events, close examination reveals as many differences as similarities. Woodstock was experienced as participatory, communitarian, and noncommercial (indeed, anticommercial), with no great (spiritual) distance between artist and audience. Interestingly, these are all terms that come from the vocabulary of folk culture. But it was Woodstock that ushered in the big-business/mass-music/technoculture of the contemporary era. To deal with the seeming irreconcilability of "folk" values with the commercial imperatives of mass culture, counterculturalists often sought refuge in the social relations of an idealized past. The hippie diaspora that was the "Woodstock Nation" thus reflected a longing for the imagined simplicity of an earlier rural life even as it embraced the electronic — not to mention the sexual — revolution.

Live Aid, by contrast, was hardly an occasion for folksy nostalgia (Baez's comments notwithstanding); it was an unabashed celebration of technological possibilities. While Woodstock was hailed as countercultural, there was precious little at Live Aid that could have been vaguely construed as "alternative," or even oppositional. If Woodstock represented an attempt to humanize the social relations of mass culture, Live Aid demonstrated the full-blown integration of popular music with the star-making machinery of the international music industry. Paradoxically, Live Aid may have opened up spaces for cultural politics that would have been unthinkable at the time of Woodstock.

Many observers have commented of late on the degree to which rock music, in its myriad forms, has become mainstream American music. "Rock & roll is now the music of the land," opined Bill Graham. "Broadway. Movies. TV commercials. *Miami Vice*. It's the music of America. It's certainly not the music of the alternative society." [1] Pronouncements such as these often serve to confirm the inevitability of co-optation and incorporation. How, then, do we explain the fact that there is scarcely a social issue in the 1980s that has not been associated in a highly visible way with popular music and musicians? Hunger and starvation in Africa, apartheid, the farm crisis, peace, political prisoners, the environment, child abuse, racism, black-on-black violence, AIDS, Central America, industrial plant closings, and homelessness have all been themes for fund-raising concerts, the subjects of popular songs, or both. (Interestingly, abortion, the issue that will probably be the deciding factor in the next round of national elections, is the one exception.) Even a cursory look at these projects reveals a liberal-to-left-leaning bias in both the choice and treatment of issues.

Chiefly responsible for this development has been the phenomenon of "mega-events"—that string of socially conscious mass concerts and all-star performances beginning with Band Aid, Live Aid, and "We Are the World" that has been dubbed, in true liberal fashion, "charity rock." While the designation itself indicates a conception that exists well within the bounds of mainstream political debate, the impact of this phenomenon represents a more serious challenge to the existing hegemony. From Bruce Springsteen's "Born in the U.S.A." to Public Enemy's "Fight the Power," the link between popular music and political issues is more explicit than ever before. There is also a significant difference in the role of popular music in building political movements. Even in the "music-and-politics" sixties, music generally served as a cultural frame for what were more or less developed political movements. The civil rights and antiwar movements engaged millions of people in the politics of direct action primarily on the strength of the issues themselves. In the process, these movements exerted a profound influence on the themes and styles of popular music.[2] In the 1980s, music—which is to say, culture—has taken the lead in the relative absence of such movements. With the decline of mass participation in grass-roots political movements, popular music itself has come to serve as a catalyst for raising issues and organizing masses of people.

This situation not only turns the traditional Marxist analysis on its head, it renders inadequate even the more sympathetic "culturalist" treatment of mass culture. Historically, our attempts at understanding mass culture have been fraught with false starts, misconceptions, and faulty analyses. Culture currently occupies center stage as a category for investigation—sometimes to a fault—but, as yet, we have no agreed-upon models or analytic tools for thinking about it. The phenomenon of mega-events represents a unique convergence of forces that challenges the very terminology of mass culture as well as our thinking about the nature of political work.

Background

Culture Theory

In the classical interpretation of Marx, society is composed of two main parts: the "base," which is the economic structure, and the "superstructure," which is the realm of values, beliefs, and ideas. Social transformation comes about as a result of radical political activity based on class contradictions in the economic realm (the base). Culture (which is part of the superstructure) is presumed to be "reflective" of ideas that are favorable to the ruling class. Because culture is "determined" in this way, it does not possess a social effectivity of its own; it is considered "nonproductive." The cultural arena, therefore, is not conceived of as a primary site for political struggle. Subsequent applications of this model to a systematic analysis of mass culture—most notably the work of the Frankfurt school—invariably yielded gloomy conclusions:

Mass culture is produced only for profit. In commodifying human interaction, mass culture reduces culture to its exchange value and negates the possibility of any real use value.

The production of mass culture is top-down and totally manipulated. The culture industry is in complete control and responses are determined in advance.

The consumption of mass culture is necessarily passive and mindless. There is no possibility for resistance/opposition within mass culture.

Is it any wonder that politically minded people avoided associations with mass culture at all costs? High culture could be appreciated for its literacy; folk culture, for its historical significance. But popular culture—that messy third tier that was itself a product of industrialization, commercialism, and the transition to capitalism—belonged to neither camp. Most often commodified and sold as mass culture, popular music could be, at best, tolerated as "entertainment" or, worse, dismissed as hopelessly reactionary. The music that played an important, if supportive, role in political movements—from labor to civil rights—was what we think of as "folk" music. But already there is a problem with terminology and, therefore, with the analysis.

There is a significant difference between the sociological use of terms like *folk, popular,* and *mass* and their meaning in everyday language. Woody Guthrie, for example, is remembered as a folk artist. But *folk* in this sense is a marketing category that was created by the music industry, in part, to separate performers like Guthrie from those producing commercial country music. In the sociological use of the term, Guthrie exhibited none of the characteristics of a folk performer. He was a known artist. He was a paid professional. He was not a member of the community he sang about. He appeared in formal settings that separated artist from audience. It would be more correct to say that Guthrie was a popular artist performing in a folkloric idiom. When his work was recorded and sold, he became a product of mass culture. Still, as late as the mid-1960s, the distinction between "authentic" folk and "commercial" pop was a significant dividing line politically. Remember that Bob Dylan was booed off the stage at the 1965 Newport Folk Festival simply for appearing with an electric guitar (his classic rock Fender Stratocaster).

When activists finally acknowledged a connection between politics and popular music in the late 1960s, the music was valued precisely when it was thought to be *something other than* mass culture or when it was deemed to be progressive *despite* its being mass culture. *Sgt. Pepper* was Art. Dylan's lyrics were Poetry. Woodstock was Community. And, Motown—well, Motown was Black. At no point was this music celebrated *as* mass culture, even though this was clearly the basis for its widespread—mass, if you will—appeal. This failure to embrace mass culture as mass culture has contributed heavily to our inability to grasp its political potential.

Interestingly, it was, in many ways, the movements of the sixties that forced a reconsideration of the traditional Marxist model at the experiential level. Participation in the major movements of the decade—civil rights, antiwar, and, later, women's liberation—typically cut across class lines. In all of these struggles, the locus of organizing as well as the major political victories were in the realm of the superstructure. Marxist orthodoxy offered little in the way of explaining these admittedly short-lived successes. The culturalist critique proposed a conception of the superstructure as "productive," in that it reproduced the social relations of production. The narrow interpretation of simple "determination" was replaced with a more dialectical process of shaping and influencing. Culture was thus accorded a certain "relative autonomy."

In the area of popular music, culturalists have tended to focus their investigations on the power of the audience to "reappropriate" culture, to determine meaning in the act of consumption. This perspective has had the positive effect of freeing human subjects from the prison of economic determinism and restoring them to their rightful place as the actors who make history. But, in concentrating to such a degree on the power of the superstructure, culturalists have also tended, often by omission, to accept the mode of production as a given. In the extreme, economic relations are overlooked as the abstract process of "resignification" magically eludes material market forces. The production of culture is no longer seen as a necessary component for thinking about political struggle.

Regarding the political potential of mega-events, this is not only a grave theoretical error, but a missed opportunity. The progressive effects of mega-events have every bit as much to do with the intentions of their artist-organizers as they do with creative consumer usage. It is in the dialectic of production and consumption that the politics of these events is realized. In exploring the power of the consumer, the culturalists have identified certain possibilities for resistance and opposition within mass culture. It is equally important to understand (and influence) the relative power and political tendencies of artists, recording companies, and the mass media on the production side of the equation.

Artist Power

In the early 1950s, the power center of the music industry shifted from an alliance of publishing houses and film studios to record companies. This shift corresponded to the ascent of records as the leading source of revenue in the business. The market was dominated by a handful of major labels, with dozens of hungry independents waiting in the wings. As per the conventional wisdom of the day, popular artists were kept on a relatively short artistic leash, as the majors controlled the production process from start to finish. All the elements—from songwriting, artist and repertoire, arrangements, production, and engineering to mastering, pressing, promotion, marketing, distribution, and, in some cases, retail sales—were organized as in-house functions. It was expected that audiences would respond favorably to gentle changes in popular styles, which would render the market that much more predictable.

The eruption of rock 'n' roll into this placid scenario demonstrated not only the relative autonomy of the cultural sphere but also the limits to which public taste could be determined from above. Far from being passively consumed, rock 'n' roll was a music that engaged its audiences—in the social ritual of dancing, in celebrating the sounds of urban life, in multicultural explorations. In resisting this new music, the majors not only acted against their economic self-interest, they contributed significantly to the oppositional posture of the form.

By the late 1960s, the majors had learned from their rock 'n' roll mistakes. As it became clear that the key to profitability lay in manufacturing and distribution, record companies began contracting out most of the creative functions of music making. Far from resisting artistic innovation or the creative impulses of independent producers, the majors simply bought up successful independent labels and artist-owned companies, entered into joint ventures with them, or contracted with them for distribution. Accordingly, among the top-selling records in the United States today, there is seldom a single entry that is not owned and/or distributed by a major label.

Like all capitalist enterprises, the music business tends toward expansion and concentration. In 1978, five transnational music corporations controlled—through ownership, licensing, and/or distribution—more than 70 percent of an international music market worth more than $10 billion.[3] Following a period of recession that plagued the industry from 1979 to 1983, revenues from the international sale of recorded music grew to $17 billion by 1987.[4] Many U.S.-based major labels now report that more than 50 percent of their income comes from sales outside the United States. At the same time, the number of new LPs released in the United States—by far the world's largest producer—declined from 4,170 in 1978 to 2,170 in 1984.[5] Far from scouring the world for new, exciting, and diverse talent—fortunately, the indies do that—major companies reap greater rewards from fewer artists.

As is the case with other culture industries, the music business is organized according to a star system. The difference is that only a handful of superstars from other cultural sectors can match the earning power of international rock stars. The entertainment press is rife with stories of sports figures and Hollywood actors who negotiate deals for $6-7 million. The top names in popular music, such as Prince, Michael Jackson, and Bruce Springsteen, take home four to five times that amount in a good year. In the United States, the revenue from the sale of recorded music alone rivals that of all organized sports and the entire film industry combined. Only television is bigger, and very few of its stars can compete with popular musicians for earnings. In addition to their considerable economic clout, top popular musicians enjoy an artistic autonomy that is unsurpassed anywhere in the cultural sphere.

While it is true that a handful of major corporations maintain a tight control over the music market, it is important, for our purposes, to note that acquiring the lion's share of the market is not synonymous with controlling the form, content, and style of popular music. If anything, record companies have relinquished this control in their relentless pursuit of higher profits. As the industry has expanded, record companies have moved further and further away from the creative process. Particularly in the 1980s, music, to these corporate giants, has come to represent a "bundle of rights." Far from being limited to the manufacture and distribution of a fixed sound product, the exploitation of "secondary rights" for things like television, movies, and advertising has become an ever-increasing source of revenue. Content to focus its energy on the development of such new sources of income for music, the industry has shown little inclination to intervene in its content (except when forced to do so by organizations like the PMRC—but that is another discussion). Artistic autonomy is virtually assured for the big names in popular music. As "Sun City" organizer Little Steven states: "These guys are in complete control of their own destinies." [6]

Technology

Popular music and musicians currently bear a different relationship to the tools of artistic production than was previously the case. Prior to the rock era, it was the function of recording to approximate the sound of a live performance. Now it's the other way around. It is the task of the live performance to reproduce what is possible in the studio. One of the reasons the Beatles stopped touring after 1966 was that, given the state of portable technology at the time, virtually none of the material from *Sgt. Pepper* could be performed live. Touring groups today routinely use "sampled" sounds and prerecorded tracks as an integral part of the music performed in concert. The question is no longer, Is it live or is it Memorex? Rather, it is, What is "live"?

Rock 'n' roll differed from previous forms of music in that records were its initial medium. They were one of its defining characteristics. Since that time, it has become more and more difficult—in many cases, downright impossible—to separate "music" from the technology used in its creation. Recording equipment can no longer be viewed simply as the machinery that reproduces something called music, that already exists independently in some finished form. Ever since magnetic tape first mediated direct-to-disk recording, successive advances in studio technology—editing, overdubbing, multitracking, and digital effects—have been used (ironically, in the name of higher "fidelity") to create products that bear little resemblance to anything that can be performed "live" in the traditional sense of the word. Particularly with the advent of digital electronics, technology exists as an element of the music itself. Music, musicians, and technology are inseparably fused in the process of creating popular music.

The cultural products that make use of this technology, therefore, cannot be dismissed as the commodities that reduce culture to an exchange value and put that much more distance between artist and audience. Recordings, worldwide broadcasts, and music videos must be seen as new forms of communication that create new modes of consumption, different, perhaps, from a "live" performance, but not automatically alienating because of that. From the time the voice was first accompanied, music making has been intimately connected to increasingly complex technologies. As was the case with earlier technologies, the music being composed on the current generation of electronic devices represents simply the most recent development in the extension of our control over the production of sound. It isn't that this music is somehow less "authentic" than other musics, it's that our feelings about authenticity—like our copyright laws and our theories of culture—have not kept pace with technological advances.

While it is true that technological advances serve the capitalist goals of expansion and concentration, it is important to note that capital itself is not monolithic in these developments. The economic self-interest of Japanese hardware manufacturers who make digital audiotape (DAT) recorders, for example, is quite at odds with that of record companies, whose primary task is to protect the "integrity" (i.e., financial viability) of the "artistic property" that is their economic lifeblood. There have been a number of "summit" meetings between these two segments of the industry to try to arrive at solutions to the problems posed by developments like DAT that would give consumers the capability of making studio-quality recordings in their living rooms. This tension has sometimes been played out among different departments of the same firm (e.g., Sony, which owns CBS Records).

Perhaps more important, capital is not the only player in the game. Technology is invariably a double-edged sword; in its development, artists and consumers also gain significant power. The same electronic advances that permit simultaneous worldwide broadcasts and the construction of international mega-audiences for capital have also encouraged decentralization in the creative process. Explains producer Niles Rogers:

> We're working out systems where if somebody who lives in
> England, say, has a system similar to what I have, and he's got a
> track and he wants me to play on it, well, he can send it to me over
> the satellite to New York. My system can then pick it up. It will go
> down on tape. I can listen to it, put my guitar overdub on it, send it
> back to him, and it'll all be digital information. It will sound exactly
> the same as when I played it. . . . I can play on your record if you're
> anywhere.[7]

In this instance, technological advances have altered the social relations of industrial production, with the result of greater and greater
degrees of artistic freedom.

On the consumption side, the electronic items that have caught the
public's fancy have been those that have delivered improved sound
quality and provided for maximum flexibility and portability of use.
In the words of Simon Frith:

> The major disruptive forces in music in this century have been new
> devices, technological breakthroughs developed by electronics
> manufacturers who have very little idea of their potential use. The
> only lesson to be learned from pop history (besides the fact that
> industry predictions are always wrong) is that the devices that
> succeed in the market are those that increase consumer control of
> their music. [8]

Particularly since the advent of digital recording technology, the
very distinction between production and consumption has become
less clear. Producers who make use of sampling devices and "found"
sounds in the creative process are, in a very real sense, acting as consumers of sounds that have been provided for them. Conversely,
home taping enthusiasts who make their own recordings by mixing
and editing different LPs and radio broadcasts are producing their
own cultural products. And, with even more sophisticated consumer
equipment just on the horizon, they will soon be able to do so with
professional quality. Once again, we are forced to rethink the conventional wisdom of theoretical categories. The advent of mass culture can no longer be seen as the historical schism that marked the
transition from active music making to passive music consumption.
Quite to the contrary, recent developments in the technology of mass
culture have transformed consumption itself into a potentially creative act.

If the cultural sphere generally is relatively autonomous, then popular music may well be its most potent sector. By virtue of the eco-

nomic power and artistic freedom of popular musicians and the particular relationships among music, musicians, and the new technology, popular music enjoys unparalleled access to the means of international communication. The political potential that accompanies this access has been played out most dramatically on the terrain of mega-events and performances.

Mega-Events

Typically, mega-events involve the creation of a variety of cultural products—live performances, worldwide broadcasts, ensemble recordings, compilation LPs, home videos, and/or *The Making of . . .* documentaries—each of which can be produced and consumed in a variety of ways.[9] It is now literally possible for hundreds of millions of people to "attend" the same concert simultaneously, be it at the "live" event, at a public broadcast, or in the privacy of their own living rooms. For those who tape a broadcast off the air or purchase a subsequent audio or video recording, the event can be relived any number of times, publicly or privately, in any number of new contexts. Audio and video recordings that illustrate a particular issue can also be used effectively as educational tools for classrooms, interest groups, and community meetings. A feedback loop is thus completed, as the consumption of the original event is used in the service of producing another. How are we to weigh the impact of all these possibilities politically?

Tony Hollingsworth, producer of the Nelson Mandela Seventieth-Birthday Tribute, outlined four functions of the Mandela concert that could apply equally well to other mega-events: fund-raising, consciousness-raising ("to raise the profile of Nelson Mandela's name as the symbol of fighting against apartheid"), artist involvement ("to demonstrate to the world and to South Africa the enormous popular and artistic support for all those who fight against apartheid"), and agitation ("to make the show act as a flagship . . . whereby the local anti-apartheid movements could pick up from the enormous coverage that we had and run a far more detailed political argument than you could have on a stage").[10] While some of these functions proved more worthwhile than others at the Mandela Tribute, they nevertheless provide useful categories for assessing the impact of mega-events in general.

Fund-Raising

Perhaps the most obvious — if not always the most important — use of mega-events is that of fund-raising. As Will Straw points out, "The most under-rated contribution rock musicians can make to politics is their money, or ways in which that money might be raised." [11] For all the reasons articulated above, popular recording artists are in a particularly good position to exploit the fund-raising potential of mega-events. "Give me another element of our society that could have drawn as many people as Live Aid," remarked promoter Bill Graham. "A sporting event? An international soccer match? I don't know. I'm trying to show what a rare position these artists are in — that a group of people can say: You want to raise $10 million?" [12]

In this regard, it is interesting to note that almost all the estimates of Live Aid's fund-raising potential were low. Not even the producers understood the power of what they were dealing with. Geldof was originally shooting for $35 million. *Newsweek* predicted that the event could "make as much as $50 million." *Rolling Stone*'s computations put the net at "more than $56 million." The actual take was $67 million. By anyone's fund-raising standards, that is a staggering amount of money to be generated from a single event.

To be sure, there are worse things that one could do with $67 million than try to feed starving people in Africa. On the other hand, $67 million is a drop in the bucket compared to the sums that are spent to create and perpetuate such problems in the first place. Given the numbers affected by starvation in Africa, $67 million averages out to about fifty cents per person. And even at the level of $67 million, there remains the question of the extent to which the event served the issue at hand versus the extent to which it served capital. At the 1986 ceremony of the British music industry's BPI Awards, Norman Tebbitt from the Thatcher administration "extolled Live Aid as a triumph of international marketing." [13] On this side of the Atlantic, Pepsi vice president John Costello said: "Live Aid demonstrates that you can quickly develop marketing events that are good for companies, artists, and the cause." [14] Mega-concerts may be uniquely capable of generating mega-sums of money, but they are equally capable of opening new markets, constructing new audiences, and delivering new consumers. Artists, record companies, and advertisers such as Pepsi, Eastman-Kodak, AT&T, and Chevrolet gained access to an international audience of one and a half billion people very cheaply because of the "humanitarian" nature of the event. Still, $67 million was raised that wouldn't otherwise have been.

As for the fund-raising function of mega-events, there are even more questions on the expenditure side of the ledger. The most obvious of these is, of course: Where does the money go? Huge concerts may be unsurpassed in their fund-raising potential, but they have not always had an exemplary track record in getting the money to its proper destination. That check that was pictured on the back of the *Concert for Bangladesh* LP, for example, languished in an IRS escrow account because of the organizer's failure to set up proper tax-exempt conduits. The quarter of a million dollars or so that the Rolling Stones raised for Nicaraguan earthquake victims in 1972 probably went straight into Somoza's pocket. By the 1980s, event organizers had gotten considerably more sophisticated about things financial. USA for Africa was set up as a tax-exempt foundation to receive and distribute the proceeds from "We Are the World." The Sun City Project chose as its conduit the Africa Fund, a tax-exempt foundation with a demonstrated track record of getting the money to where it is supposed to go.

The more pressing question about fund-raising is: What does the money get used for? In the aftermath of the concert, Live Aid was beset by reports of food rotting on docks and trucks that didn't work— charges that Geldof refuted repeatedly as inaccurate. But we are not just talking about the mechanical problems of implementing any large-scale project. We must also raise the question in a broader philosophical context. In thinking about fund-raising politically, it is important to distinguish between charity and change, dependence and self-determination, quick fixes and long-term development.

To their credit, these are distinctions that many artists and organizers of mega-events have at least thought about. While the character of Live Aid and certainly most of the reportage about the event were decidedly apolitical, the Band Aid Trust did entertain proposals for "the purchase of water-drilling rigs to help with irrigation; various agricultural projects, including reforestation; medical aid; and the purchase of trucks and trailers for transportation of food and supplies." [15] Still, significant relief has never materialized, because Geldof and his organization failed to account adequately for the difficulties associated with the ongoing civil war in the region. The Africa Fund distributed the proceeds from "Sun City" in more or less equal shares to "political prisoners and their families . . . in South Africa; . . . the educational centers and college set up by the ANC in Tanzania and Zambia; . . . grassroots educational outreach by the anti-apartheid movement in the US." [16] In the case of the Amnesty International tours, the proceeds went to fund the ongoing political work of the organization.

It must be noted, however, that the operations of the mass media make it more difficult to get the public to think about the long haul. "Long-term aid is less exciting than the Seventh Cavalry arriving with food to bring people back to life," said Geldof. "And that's a problem." [17] Further, the crisis orientation of the media exacerbates a more general problem with fund-raising as a strategy—namely, the idea that simply generating huge sums of money can solve problems that are fundamentally political. This is why the consciousness-raising function of mega-events is so important.

Consciousness-Raising

If the power of popular music readily lends itself to fund-raising, then its form is equally well suited to consciousness-raising. It is laudable that John MacEnroe has refused million-dollar deals to play tennis in South Africa on more than one occasion. But there is no way that tennis as a cultural form can portray the horrors of apartheid. While any cultural event can be dedicated to a particular cause, popular music is further distinguished in its ability to reflect the issue at hand in its very content. Of course, other cultural forms such as film or theater have this capability, but they seldom combine the versatility, responsiveness, and impact of popular music.

It is always tempting—and, more often than not, too facile—to write off projects like USA for Africa. Greil Marcus, for example, has argued that "We Are the World"

> sounds like a Pepsi jingle—and the constant repetition of "There's a
> choice we're making" conflates with Pepsi's trademarked "The
> choice of a new generation" in a way that, on the part of Pepsi-
> contracted song writers Michael Jackson and Lionel Richie, is
> certainly not intentional, and even more certainly beyond the realm

of serendipity. In the realm of contextualization, "We Are the
World" says less about Ethiopia than it does about Pepsi—and the
true result will likely be less that certain Ethiopian individuals will
live, or anyway live a bit longer than they otherwise would have,
than that Pepsi will get the catch phrase of its advertising campaign
sung for free by Ray Charles, Stevie Wonder, Bruce Springsteen, and
all the rest.[18]

There was perhaps an even more distasteful element of self-indul-
gence in the follow-up line "We're saving our own lives," where the
artists assembled proclaimed their own salvation for singing about
an issue they will never experience on behalf of people most of them
will never encounter. With hype, glitter, and industry gossip often taking precedence over education, analysis, and action, Live Aid was also
vulnerable to any number of criticisms.

At the same time, we must not underestimate the political importance of the momentum these projects generated. In the first place,
they effected an international focus on Africa that was simply unprecedented. In the process, they created a climate in which musicians
from countries all over the world felt compelled to follow suit. A partial list of African famine relief music projects is sufficient to illustrate
the point:

> Great Britain: Band Aid, "Do They Know It's Christmas?" (37 artists)
>
> United States: USA for Africa, "We Are the World" (37 artists)
>
> Canada: Northern Lights, "Tears Are Not Enough"
>
> West Germany: Band für Ethiopia, "Nackt im Wind"
>
> France: Chanteurs Sans Frontières, "Ethiopie" (36 artists)
>
> Belgium: "Leven Zonder Honger"
>
> The Netherlands: "Samen"
>
> Australia: "E.A.T." (East African Tragedy)
>
> Africa: "Tam Tam Pour L'Ethiopie" (50 African artists, including
> Youssou N'Dour, Hugh Masakela, Manu Dibangu, and King Sunny
> Ade) [19]

Even the most cautious humanitarian efforts can create the cultural
space for bolder undertakings. Just as Live Aid begat Farm Aid, the
whole "charity rock" phenomenon has inspired other, more politicized ventures, like Amnesty International's Conspiracy of Hope and
Human Rights Now tours. The Mandela Tribute would have been unthinkable without Live Aid. The focus on Africa, which began with the

relatively safe issue of hunger, quickly targeted the more compelling issue of apartheid. Using "We Are the World" as a model, Little Steven assembled more than fifty rock, rap, rhythm and blues, jazz, and salsa artists to create "Sun City," a politically charged anthem in support of the cultural boycott of South Africa.

An artist's involvement with political issues and events can also be the occasion for the emergence of a more politicized popular music. In the words of Simon Frith and John Street: "The paradox of Live Aid was that while in the name of 'humanity' it seemed to depoliticise famine, in the same terms, in the name of 'humanity' it politicised mass music." [20] Following the example of "Sun City," Stevie Wonder released "It's Wrong (Apartheid)" and dedicated his 1986 Grammy to Nelson Mandela. Kashif took on the issue of apartheid with "Botha, Botha." Jim Kerr of Simple Minds wrote "Mandela Day" especially for the Mandela Tribute. Stetsasonic delivered the searing rap and video "A.F.R.I.C.A."

Politicized popular music has extended to a broad range of other issues as well. There is a clear connection, for example, between John Cougar Mellencamp's involvement in Farm Aid and his "Rain on the Scarecrow," a song about the despair of modern rural life. Jackson Browne's interest in Central America led to "Lives in the Balance," a moving criticism of U.S. intervention in Central America. A number of rap groups, including Public Enemy, Boogie Down Productions, and Stetsasonic, participated in "Self-Destruction," the anthem of the Stop the Violence movement, protesting black-on-black crime. In recording "Fight the Power" for Spike Lee's *Do the Right Thing,* Public Enemy contributed the soundtrack for the most powerful statement about racism in recent memory. As more and more songs born of political experience enter the popular market, the development of a more politicized culture gets validated.

Early in 1989, Geffen Records and cable music channel VH-1 teamed up to promote a mega-project to benefit Greenpeace. As their part of the project, VH-1 produced more than two dozen sixty-second spots, called "World Alerts," that featured celebrities discussing a range of environmental issues. Additionally, artists ranging from U2 and Talking Heads to John Cougar Mellencamp and Belinda Carlisle donated twenty-seven hit songs to a compilation album titled *Rainbow Warriors.* The same double LP was also released in the Soviet Union as *Breakthrough.* It soon became the top-selling record in the USSR, with all proceeds split between Greenpeace and the Foundation for Survival and Development of Humanity. A few months later, a number of Western heavy metal acts, including Ozzie Osborne, Mötley Crüe, and Bon Jovi, participated in the Moscow Peace Festival, the first Soviet mega-festival to be broadcast worldwide. Popular music has forced a level of cross-cultural communication that governments have resisted for years.

Our Common Future, another environmental extravaganza, was staged at Lincoln Center in the spring of 1989. Participating were Bob Geldof, Richard Gere, Sting, Midnight Oil, and Herbie Hancock, among others. In addition to top-notch entertainment, the show provided a platform for a number of scientists and world leaders to voice concern over global environmental decline. It was a little unnerving, however, to see Margaret Thatcher delivering a pretaped message about Britain's concern for the environment. It may be that the politics of the show were complicated by the sponsorship of multinational corporations, including Sony, Panasonic, and Honda, all of whom had their corporate logos prominently displayed during the syndicated telecast.

Dave Marsh in particular has been critical of corporate involvement in mega-events for robbing "charity-rock of one of its most important selling points: the selflessness of its motivation." [21] Given the scale of mega-events, however, most would be impossible without some kind of corporate involvement. Amnesty International's Human Rights Now tour would have gone bankrupt had not Reebok bailed them out at the last minute. Further, there is a more optimistic reading of the situation: that the power of popular music—and, in particular, the phenomenon of mega-events—has obliged corporations and world leaders to accommodate initiatives that are essentially left wing or progressive.

Artist Involvement

Celebrity endorsement has long been used to bolster campaigns, support charities, and sell products. Why not to promote social causes? In this regard, it is important to note that mega-events and socially conscious mass music have been the beneficiaries of a left-leaning orientation that has characterized popular music since the rock era. As explained by New York Times critic John Rockwell:

Rock's leftist bias arose from its origins as a music by outsiders—by blacks in a white society, by rural whites in a rapidly urbanizing economy, by regional performers in a pop-music industry dominated by New York, by youth lashing out against the settled assumptions of pre-rock pop-music professionals.

That bias was solidified by the 1960's, with its plethora of causes and concerns. . . . Rock music was the anthem of that change— racial with the civil-rights movement, and also social, sexual, and political. [22]

For the most part, popular music still seems to draw from that sixties spirit, but as the music becomes more and more mainstream, there is no guarantee that this will remain the case. Lee Atwater, George Bush's campaign manager, managed to co-opt blues and soul artists such as Joe Cocker, Albert Collins, Steve Cropper, Bo Diddley, Willie Dixon, Dr. John, "Duck" Dunn, Sam Moore (of Sam and Dave), Billy Preston, Percy Sledge, Koko Taylor, Carla Thomas, Stevie Ray Vaughn, and Ron Wood, among others, for an inaugural performance for Bush. White supremacist organizations have discovered the power of punk in recruiting skinheads to their cause. In these politically conservative and economically uncertain times there are any number of forces that eat away at the progressive edge of the music and complicate the nature of artistic involvement.

In producing mega-events, there is an inevitable tension caused by choices made in recruiting "name" artists who will ensure the financial success of the event versus local artists or artists who have a demonstrated commitment to the issue at hand. From another angle one has to wonder what motivates an artist to become involved—political commitment, economic considerations, public relations, mock heroism, ego? Different events require different levels of political commitment. "We Are the World" was recorded essentially in one session on the night of the 1985 Grammy Awards ceremony, when all of the contributing artists were already in Los Angeles. Similarly, Live Aid demanded only one day of the artists' time and no particular political commitment to anything beyond some basic notion that starvation is a bad thing. Still, Bob Geldof did not hesitate to resort to what was referred to as "moral blackmail" in recruiting artists. With twenty-one dates in eighteen countries, the Amnesty International Human Rights Now tour made far greater time demands on its artists. Still, headliners Bruce Springsteen, Peter Gabriel, Sting, Youssou N'Dour, and

Tracy Chapman not only made all the dates, but also participated in press conference after press conference.

To the extent that exposure is the name of the media game, one must question an artist's motivation for performing at a live mega-event. Even though artists technically play for free at most of these concerts, when one considers the impact of having an audience of hundreds of millions in terms of, say, record sales over the next few days, then every artist who performs probably has one of the biggest paydays of his or her career. Questioned along these lines regarding his appearance at Farm Aid, Billy Joel retorted: "We don't need exposure." [23] To be fair, it must be noted that overexposure is just as big a concern for many of the artists who headline mega-events.

There are times when an artist's personal politics can be perceived as having a direct effect on the political character of an event. Peter Jenner, Billy Bragg's manager and longtime political activist, was critical of the Mandela Tribute for booking Whitney Houston. "The story around was that she would not do the show unless it was non-political," said Jenner. "The moment they agreed to those terms, they lost the battle." [24] Jenner argued further that the move offended lesser-known artists with a track record of antiapartheid performances, like Bragg, who were passed over for bigger stars. "They're now reluctant to do any antiapartheid events," he added.[25]

On the other hand, Houston delivered easily the most animated performance of the day. While it is true that her management made a "sharp distinction" between "what is humanitarian grounds and what is politics," it is also the case that she was one of the first acts to commit to the festival.[26] Her overwhelming popularity contributed significantly to making a controversial event that much more attractive to broadcasters all over the world. And her cautious politics did not seem to bother the ANC. Ahmed Kathrada, one of the ANC rebels who received a life sentence along with Mandela, sent a message that was distributed by the local Anti-Apartheid Movement. "You lucky guys," wrote Kathrada from his cell. "What I wouldn't give just to listen to Whitney Houston! I must have told you that she has long been mine and Walter's [Sisulu] top favorite. . . . In our love and admiration for Whitney we are prepared to be second to none!" [27]

Hollingsworth defended his choice on the grounds that it was the idea of a musical tribute that defined the politics of the Mandela concert and that Houston "agreed to pay tribute to Nelson Mandela as the symbol of fighting against apartheid." [28] Nevertheless, the degree of "fit" between artist and issue is one variable that can affect the public perception of a mega-event. The relative absence of black acts at Live Aid (and the poor coverage of some of those who were there), for example, was especially noticeable given the nature of the issue and the fact that black artists had provided all of the leadership for "We Are the World." A less obvious, but equally important, contradiction was the presence of performers such as the Beach Boys, Queen, Tina Turner, and Cher, all of whom had played South Africa in spite of the U.N.-sponsored cultural boycott.

On the positive side, many of the artists who headline mega-events have donated their time and talent to local communities and political organizations. Little Steven immersed himself in the issue of South Africa for the "Sun City" project. Following that, he turned his attention to organizing in the Native American community. Because of his involvement in Central American issues, Jackson Browne did a series of benefit dates for the Christic Institute. Bruce Springsteen rallied when 3M proposed closing its plant in Freehold, New Jersey. All of the profits (including those of the artists, record company, producers, publishers, and participating unions) from Dionne Warwick's "That's What Friends Are For," which also featured Elton John, Gladys Knight, and Stevie Wonder, were donated to the American Foundation for AIDS Research. The rap artists who initiated the Stop the Violence movement have played an ongoing role in protesting the explosive conditions that exist in communities of color.

Agitation

The intentions of artists and producers clearly exert a profound influence in shaping the political character of our popular culture. At the same time, cultural products may have unintended consequences as well. Who, for example, would have predicted that striking black South African students would be chanting, "We don't want no education. We don't want no thought control," lines from Pink Floyd's "Another Brick in the Wall," as they boycotted schools? In order to assess the impact of a cultural phenomenon, it is essential to look at how it gets used.

Tony Hollingsworth envisioned the Mandela Tribute as a "flagship event" to be used by local antiapartheid groups. In England, the site of the live event, the concert was used to advantage by the local Anti-Apartheid Movement to enlist support for its "Nelson Mandela: Freedom at Seventy" campaign. According to Hollingsworth, the resulting momentum forced a change in the media coverage of Mandela and the ANC. In Rome, the festival was sponsored by *Il Manifesto,* an independent left-wing newspaper, and was broadcast on public television channel 3, the communist channel. In the Piazza Farnese, a historic outdoor plaza, the concert was projected on a ten-by-fifteen-foot television screen to an audience of thousands for free. A small stage served as a platform for antiapartheid speeches by Italian and African political leaders. The mega-festival was consciously used to create a local political event.

The antiapartheid movement in the United States did not make direct use of the Mandela Tribute. Here, the concert was televised nationally by Rupert Murdoch's Fox Television Network as *Freedomfest,* a five-hour edited broadcast, which was widely criticized for having depoliticized the event. Nonetheless, there is a progressive side effect of mega-events like Live Aid and the Mandela Tribute and the all-star performances of "We Are the World" and "Sun City." In the United States, radio formats are designed to cater to the tastes of a fragmented audience. Ostensibly these divisions represent differences in musical preferences, but, conveniently, they correlate highly with divisions of class, race, age, and ethnicity. The artists who participated in the mega-performances mentioned above encompassed a broad range of audience demographics and radio formats. The media outlets that carried these performances contributed significantly to breaking down the apartheid of our own music industry. "Whoever buys [the *Sun City* LP]," remarked coproducer Arthur Baker, "is going to be turned on to a new form of music, just as whoever sees the video is going to be turned on to an artist they've never seen before." [29]

"Sun City" also broke new political ground in its attempt to encourage an activist audience response. The *Sun City* album jacket, for example, was filled with facts and figures about apartheid. In addition, the Sun City Project issued a "Teacher's Guide" that showed how to use the record and the video as educational tools in the classroom. As part of this educational effort, the "Teacher's Guide" reported on numerous antiapartheid student projects from all over the country that had been inspired by the "Sun City" recording. Here the attempt was made to build on the familiarity of the mass-cultural product to create exercises that could be tailored to local use.

Amnesty International also encouraged political activism in the way its tours were organized. The Conspiracy of Hope tour targeted six political prisoners as part of the event. One of the goals was to recruit new "freedom writers" who would participate in the letter-writing campaigns Amnesty International uses to call attention to the plight of prisoners of conscience. As a result of their efforts, three of the prisoners were freed within two years. In addition, Amnesty/USA added some 200,000 new volunteers to the organization. "Previous to 1986, we were an organization post forty," said Executive Director Jack Healy. "Music allowed us to change the very nature of our membership." [30] The mass cultural events were used to enlist people directly into the ongoing political activity of the organization.

Conclusion

It is likely that mega-events will continue to happen, if for no other reason than our utter fascination with their technological possibilities. But, beneath the gleaming surfaces of the pop scene, mass culture exists as a site of contested terrain. It is in this fertile arena, with all of its contradictions, that progressive forces must either make their voices heard or risk being relegated to the margins of techno-culture. Thus far, mega-events have been staged, for the most part, in support of reasonably progressive causes. Indeed, in most instances, they have shifted debate to the left. But, as Bob Geldof cautioned those who envisaged Live Aid as a sixties-style movement:

> We've used the spurious glamour of pop music to draw attention to a situation, and we've overloaded the thing with symbolism to make it reach people. But people get bored easily. People may have been profoundly affected by the Live Aid day—some were shattered by it—but that does not translate into a massive change in consciousness. [31]

Geldof's case may have been a bit overstated. It has been demonstrated that popular music is capable of far more than "spurious glamour." At the same time, we would do well to acknowledge its limitations. While one hesitates to borrow metaphors from Keynesian economics in this day and age, it would, perhaps, be fair to say that mega-events appear to be quite useful for priming the political pump. But, for those interested in lasting structural change, it has to be recognized that they are no substitute for a political movement.

NOTES

1. Michael Goldberg, "Bill Graham: The Rolling Stone Interview," *Rolling Stone* (December 19, 1985), 138.

2. For a more developed discussion of this process as it relates to the civil rights movement, see Reebee Garofalo, "The Impact of the Civil Rights Movement on Popular Music," *Radical America*, 21 (November-December 1987) (mailed March 1989), 15–22.

3. Laurence Kenneth Shore, "The Crossroads of Business and Music: A Study of the Music Industry in the United States and Internationally," unpublished doctoral dissertation, Stanford University (1983), 248.

4. *IFPI Newsletter* (January/February 1989), 3.

5. *Inside the Recording Industry: A Statistical Overview, Update '86* (New York Recording Industry Association of America, 1986), 5.

6. Personal interview with Little Steven, New York (July 18, 1989).

7. Ted Fox, *In the Groove* (New York: St. Martin's, 1986), 334–36.

8. Simon Frith, ed. *Facing the Music* (New York: Pantheon, 1988), 129.

9. A few passages from this section have appeared in somewhat different form in Reebee Garofalo, "Nelson Mandela, The Concert: Mass Culture as Contested Terrain," ed. Mark O'Brien and Craig Little, *Reimaging America: The Arts of Social Change* (Philadelphia: New Society, 1990).

10. Tony Hollingsworth, remarks from "Mass Concerts/Mass Consciousness: The Politics of Mega-Events," panel presentation at the New Music Seminar, New York (July 17, 1989).

11. Will Straw, remarks from "Rock for Ethiopia," panel presentation at the Third International Conference on Popular Music Studies, Montreal, Canada (July, 1985), 28.

12. Michael Goldberg, "Bill Graham: The Rolling Stone Interview," *Rolling Stone* (December 19, 1985), 169.

13. Simon Frith, "Crappy Birthday to Punk," *In These Times* (April 23–29, 1986), 20.

14. *Rock & Roll Confidential* (September 1985), 3.

15. Michael Goldberg, "Live Aid Take May Hit $60 million," *Rolling Stone* (August 29, 1985), 19.

16. Correspondence from Jennifer Davis, executive secretary of the Africa Fund, to Artists United Against Apartheid (September 10, 1985).

17. David Breskin, "Bob Geldof: The Rolling Stone Interview," *Rolling Stone* (December 5, 1985), 67.

18. Griel Marcus, remarks from "Rock for Ethiopia," panel presentation at the Third International Conference on Popular Music Studies, Montreal, Canada (July, 1985), 17.

19. Stan Rijven, remarks from "Rock for Ethiopia," panel presentation at the Third International Conference on Popular Music Studies, Montreal, Canada (July 1985), 3–7.

20. Simon Frith and John Street, "Party Music," *Marxism Today* (June 1986), 29.

21. *Rock & Roll Confidential* (September 1985), 1.

22. John Rockwell, "Leftist Causes? Rock Seconds Those Emotions," *New York Times* (December 11, 1988), 23.

23. "Concert Aid: Music to the Rescue," *Billboard* (December 28, 1985), T-7.

24. Personal interview with Peter Jenner, New York (July 19, 1988).

25. Ibid.

26. Telephone interview with Tony Hollingsworth (September 15, 1988).

27. Anthony De Curtis, "Rock & Roll Politics: Did the Nelson Mandela Tribute Make Its Point?" *Rolling Stone* (August 11, 1988), 34.

28. Hollingsworth, telephone interview.

29. *The Making of "Sun City,"* Karl-Lorimar Home Video (1985).

30. Jack Healy, remarks from "Mass Concerts/Mass Consciousness: The Politics of Mega-Events," panel presentation at the New Music Seminar, New York (July 17, 1989).

31. Breskin, "Bob Geldof," 33.

Black Box S-Thetix:
Labor, Research, and Survival
in the He[Art] of the Beast
Jim Pomeroy

Given the opportunity to work with new, evolving media in the "big science" [1] orientation of our fin de siècle decades, artists have rushed and hustled to take advantage of the increasingly rich technological stockpile of tools and palettes. Honors often go to those who are among "the first on the block" to pioneer aesthetic experiments with "state of the art" technology. The story of Nam June Paik's achievement as the first artist to get his hands on a Sony PortaPak VTR is just one apocryphal milestone in the chronicle of first ascents. Many a subsequent career has been underwritten by an artist's intimacy with the tides of technological innovation. Laurie Anderson's impressive performance persona, for example, is doubtless enhanced by her use of home-brewed high-tech, such as magnetic tape-bowed violins, vocal harmonizers, all-body microphones, and multi-media projections. So too, high profile use of corporate-scale media exemplifies the potential for enhanced political expression in the latest versions of our "society of the spectacle." Witness Jenny Holzer's Wall Street and Times Square screenings of the political work of eighty artists on a mobile DiamondVision truck during the closing days of the 1984 presidential campaign — an outstanding mobilization of technology to rupture provocatively the informational surround.

All across the First World, artists are experimenting with new media, exotic technology, and inventive modes of access and avenues of distribution. While the marketed range of sophisticated devices available is staggering, aesthetics is hardly the motivating force behind most high-technology research and development. In fact, much of our art is dependent on the spin-off utility generated by solutions to such artful problems as target acquisition (infrared and ultrasonic sensing, thermal imaging, radar, sonar, and microwave, stroboscopic photography, gyrostabilized target designators or steadycams, sound analysis by acoustic signature, signal processing, and image intensification), covert penetration (high-strength fiber laminates, terrain-following mapping, and inertial guidance), secure battlefield communications and management (cellular phones, FM synthesis, voice recognition, heads-up display, expert systems), pinpoint delivery of nuclear payloads (silicon chips, microcomputers, software and telecommunications, new metallurgy and ceramics), smart bombs and surveillance drones (robotics, lasers, high-resolution, low-light video), high-altitude reconnaissance (enhanced photographic emulsions and optics), and combat simulators (3D modeling, ray-traced animation, visual data bases, and virtual space), to name only a few.

While techno-art may appear to be new, the technology has, in most cases, been around for a long time. There is often a substantial lag (known as "first-flight-to-service-entry" in U.S. Department of Defense jargon) from initial R&D to its eventual use by artists because of security restrictions, the high cost of custom or prototype components, and the lack of general knowledge about likely or possible applications. Cultural integration of esoteric technologies can take as long as thirty years. A good example is the recent art world interest in digital photography and still video, derived from the imaging technologies initially developed in the mid-1960s for satellite surveillance and planetary probe explorations.

Our most common initiation into the world of innovative high-tech art is usually through highly touted presentations at world fairs, the Olympic Games, trade shows, and sundry other spectacular "bread and circus" events. Just as George Frederick Handel was commissioned to write his *Royal Fireworks Music* for eighteenth-century royal functions, Jean Michel Jarre overwhelmed the Houston skyline with lasers, pyrotechnics, and multichannel FM broadcasts in 1986's Houston Festival, while New York Harbor's Statue of Liberty makeover was celebrated by genetically recombining 200 Elvis clones live on cablevision. Corporate and government sponsorship of high-tech spectacle is well served by promoting a posture of forward-looking, sophisticated, and intimidating technical superiority. Public art presented under such auspices is usually rich in contentless abstraction or saturated in humanistic sentimentality, like the Smithsonian Air and Space Museum's "Spirit of Flight." The IBM-sponsored exhibition "World of Fractals" is an excellent example of such "art" discovered in the remote interstices of "profound" mathematical analysis. That the microscopic scrutiny and arbitrary coloration of "chaotic" numerical data are now presented as art testifies to a begrudging acceptance of abstract painting by the lay and technical communities, albeit more than fifty years after its American debut. That these seemingly complex patterns result from formulaic derivation, rather than from voluntary human activity, positions them, ironically, as the ultimate in abstract art. Mounting and touring this show serves to cloak the manufacturers of International Business Machines with the appearance of largess and cultural sensitivity, while excluding the messy involvement of artists' sensibilities, intentions, and expressive politics.

Artists who cater to the sanitized decorum of corporately acceptable formalism risk being characterized as sellouts, hacks, and mercenaries. But there are many who eagerly seek the limelight. The officially sanctioned artist's moral position in regard to the source or motivation of high technologies is at best neutral and indifferent. Out of either ignorance or agreement, they often lack the recognition that their work is used to celebrate the brokerage of frequently highly questionable power structures. The late shah of Iran was all too fond of such megabuck technotainment, to the subsequent embarrassment of not a few Soho art luminaries who were left holding the bag when the Peacock throne was vacated. By contrast, many contemporary artists independently involved in developing these technogenres acknowledge a profound ambivalence about the corporate-military origins and intended applications of the tools they use, and that attitude is often strongly reflected in the content of their work. However, this comparison says more about the complicit nature of officially sanctioned culture than it does about my intended topic—the politics of maverick survivalism in a domain bridged between the laissez-faire art market and the firmly entrenched defense-industry cartel.

While many contemporary artists employ new technologies to produce their work, most have little or no grasp of the unique technical properties, programming requirements, or fabrication demands of these tools and processes. In this essay, I will be focusing on a few of those artists who choose to make interventions within or upon the technological environments of their work, that is, artists, who, as inventors, programmers, tinkerers, pirates, or hijackers, possess the wherewithal and the skills to "hack," "customize," and "tweak" by direct, personalized, and expedient means. Inventors are classic American types, from Franklin and Edison to Rube Goldberg and Tom Swift, and their counterparts in art today exhibit new qualities, in keeping with new times. We no longer inhabit a frontier republic, distant from the urban glut and high culture imprimatur of Europe. We have imported or fabricated our own encroaching density, cultural hierarchies, and consumerist economies. The church, monarchies, and aristocracies that our rebellious forefathers rejected have been replaced by corporate hegemony and wealth reinforced through capital gains tax exemptions and the permanent arms market. The profit-motivated entrepreneurial curiosity of the nineteenth-century inventors is retooled by contemporary techno-artists as intellectual disengagement from the dominant culture, referential resonance with postmodern discourse, and the "situational aesthetics" of experimental, site-specific, and/or interventional activities.

Their schooling is different, too. Many of these baby-boom bricoleurs watched *Mr. Wizard* during the gray-flannel, silent twilit zones of the complacent, conformist 1950s, tuning up the hot rod and listening for the beatnik of a different bongo. Few were tuned in to hear Eisenhower's parting warning about the military-industrial complex. Crew-cut and slide-ruled, they were primed to respond to the rallying muster of pumped-up science education in the aftermath of *Sputnik*. Their science fair projects frequently led to college scholarships, where universities welcomed freshman engineers and physicists with crisp new labs, massively humming banks of punch-card-fed computers, and sanctuarial draft deferments. Their positivist idealism shattered by televised U.S. excesses in Southeast Asia, electronic tripping and video play in the Global Village beckoned the new Connecticut Yankee to patch together a Whole Earth of appropriate technology in the postsixties, postfeminist ecotopia (monkey-wrench pranksterism notwithstanding). Pluralist sympathies, populist consciousness, and the egalitarian technology of copier/offset, video, and performance art challenged the auratic economy of the big-time culture establishment. The art market's clumsy attempt to maintain business as usual recycled the same old notions of patronage and presentation — museum/corporate programs such as "Experiments in Art and Technology" were pretentious deadends, while collective activists and maverick upstarts invented new venues for new audiences watching new ideas. Outside the decorous parameters drawn by slick, button-down front-office art administrators, breakthrough events were being staged; in 1974, in a signal incident, San Francisco's Antfarm barreled a vid-eyed cyclopean Cadillac through a wall of flaming televisions in *Media Burn*. Recent recruits are drawn from the new wave of industrial-culture cyberpunks — the first wholly digital generation nurtured on Apples and Casios, VCRs and microwaves, MTV and CompuServe, computer viruses and HIV.

Three decades after the United States scrambled to play catch-up to the Russians, technoculture in the art community is widespread and diversely represented in science centers such as San Francisco's Exploratorium, annual experimental expos such as New Music America, artist spaces such as Buffalo's Hallwalls/CEPA and Seattle's Soundwork, media centers such as Santa Fe's Center for Contemporary Art, schools such as CalArts, UCSD, and NYC's Visual Arts, and innovative museums such as MIT's (naturally) List Center. But techno-art is also liable to show up anywhere: Chico MacMurtry's teleoperated steel mannequins walking down the street, waiting at Dennis Adams's confrontational *Bus Shelters,* under Sheldon Brown's *Video Wind Chimes,* hearing Max Neuhaus's soundworks hum up from the subway grating, beneath Rockne Krebs's trespassing laser projections of Robert Mapplethorpe's portrait thrown upon the wall of the Corcoran Gallery, floating past the drifting anthropomorphic modules of Brian Rogers's *Odyssetron* cybernauts, or an unemployed Los Angeles truck driver sailing up and off into the sunset in his lawn chair suspended beneath a huge bouquet of helium-inflated surplus weather balloons.

The new techno-art implies a folkloric familiarity with the technological processes of an advanced electronic-industrialized society. It is not a celebratory cargo cult, however, although its oppositional use is often a mirror image of emplaced technology in the defense and corporate sectors. The by-product of surplus, or trickle-down technology from those sectors, this "aesthetic research" often takes on the piratical aspect of secondhand R&D. Although the hardware may not be fresh, its implementation is frequently beyond the scope of the original designers' intentions.

New corporate technologies are almost never introduced with challenging rhetoric or confrontational aesthetics. Most premier demonstrations, big budgets and publicity notwithstanding, rely on familiar presentations—endless variations on the *Mona Lisa,* Abel's *Sexy Robots,* Pixar's sentimental, super-Disney *Luxo, Jr.* animations, and the safe brand of Hollywood Cubism and Wall Street Decoratif that is most often reproduced in the glossy coffee-table anthologies of Computer Art. Major companies rely upon programmers and illustrators rather than commissioning independently creative artists. The annual computer graphics showcase, SIGGRAPH, abounds with glitzy, vapid displays promoting the latest hard-and-soft wares out on the main floor. Many artists attend just to check out the new toys, reading between the lines of lowest-common-denominator sales pitches for a sense of the real possibilities of these gadgets. (Once a product is out long enough to debug the beta releases and stabilize at a reasonable discount price, then the boxes and source code can be kludged, cannibalized, perverted, and cross-bred.) But the real action, and the cognoscenti, is elsewhere. Corporations may tolerate freaks and nerds in the programmer pool, but they are rarely visible in the front office or on the sales floor. Behind the big business trade show, the real conference is being conducted in shop-talk critiques strung out along the endless halls connecting meeting rooms and johns. The most interesting work can be seen in the "back-room" Art Shows, generated out of slightly older systems from accessible media labs, university research associates, and the laborious result of solitary individuals tweaking hybrid micros and custom designs. The technology may be a little outdated, but it is personal, understandable, and affordable.

A similar ghettoization exists in the academy, where universities, seduced by massive government subsidies, will categorically fund most science and engineering ventures from accredited academic entities (read "lobbies"), no matter how remote or socially abstract the grant proposals sound, while reluctantly supporting humanistic projects at a fraction of the "applied" research budgets. For example, conservative estimates for the cost of the Super Conducting Super Collider ($7.9 billion for the seven-year project, November 1989 estimate) dwarf the combined totals of the National Endowments for the Arts and Humanities budgets (and may easily drain off those beleaguered appropriations or displace campus matching funds). Internal ivory tower prejudice reinforces the science/humanities split that C. P. Snow first described in 1954, in *The Two Cultures and the Scientific Revolution.* Not surprisingly, techno-artists often find a swifter stream of trickles on the nerdier side of campus than in the fine arts sector, as well as more genuine interest in their work. When a colleague photographer sought to develop a computer lab for digital imaging in the Art Department of a major West Coast university in 1986, he was stalled until he approached the Engineering College for help through the major corporate grant they had just been awarded. Shortly after, he discovered that the senior art faculty had secretly resolved to bar computers from the department curriculum. Clearly, they felt that traditional art pedagogy was threatened by visual electronics, viewed as a violation of sacred craft and media, and as an intrusion from the alien realm of science and industry. This Luddite paranoia can be seen as the academic reciprocal to the general corporate distrust of bohemian artists, egghead intellectuals, and other aesthetic court jesters. Such mutually exclusive attitudes further attenuate the narrow landscape for artists working within techno-culture.

Technological art is even less likely to fulfill the aesthetes' divine regard for "timeless" art, since a good deal of the art produced with advanced tools can become obsolete quite quickly. It will often wear out, literally, and can quickly exhaust its supply of replacement components, machinable repairs, or service knowledge. Or its novelty will fade in the wake of newer, glitzier toys rolling off the assembly line. Gadgetry for its own sake won't appeal for long—holography has been hyped for thirty years, but few substantial examples stick around long enough to maintain viewer interest, while the serious artists in the field find their credibility diluted by hyperfatigue. As described in Stewart Brand's *Media Lab,* MIT's highly touted research facility chose to demonstrate its ground-breaking work on free-floating holograms with hovering suspensions of—guess what—the newest GM sports cars! Intelligent and accessible applications take a backseat to ever-fresher tributes to corporate mystification on the part of commercial illustrator/programmers. Experiences such as these lend themselves to a skepticism bordering on the technophobic, an attitude shared by many critics, museums, and collectors (and eventually funders), and countered only by the persistence of the maverick

technogrunts slugging away in the trenches of techno-artistic practice.

Unlike salon-oriented, well-bred fine art, contemporary techno-art
usually has the look of the makeshift prototype, self-consciously "ma-
chined" and resembling something more like studio furniture, laboratory apparatus, or grease-encrusted "Road Warrior-ware" than
framed, polished Fine Art elegance. In contrast to the remote, exclusive aura of tasteful connoisseurship, techno-art is usually directly engaging and context specific. For example, New York sonic artist Liz
Phillips's straightforward, lean installations are subtly delineated by
circuitry and sensors, while the actual "work" resides only in the
space, light, and sound produced interactively on site, and must be
entered and investigated in order to elicit the intended perceptual
experience. Similarly, the striking images that pop out from Bill Bell's
Lightsticks must be readable—the effect of perceiving graphic images "drawn" across one's retina during head-eye movement is too
transient to be noticed otherwise.[2] Nonrecognizable imagery would
be lost in the background visual noise. While a degree of abstraction
may be manifest in the surface effects generated by work such as that
of Phillips and Bell, the installations are usually planned as a psychologically or socially engaging encounter for an audience. This is particularly true of the five artists whose work I have chosen to discuss
in more detail below.

Alan Rath

I moved to Oakland to be closer to the detritus of Silicon Valley.
—Alan Rath, *Digital Photography,* 1988

Recombinant elements hard wired from fragments of video games,
radar components, ancient computers, mill-end stock overages, obsolete redesigns, raw chassis, and trailing umbilicals are all part of
Alan Rath's literally deconstructed work. The 1988 *Ambivalent Desire*
is typical of his installations, which often feature the guts of monitors
extracted, stretched, exposed, and suspended into exotic diagram-
matic dissections of nervous systems and circulatory metaphors. Dis-
playing animated fragments of photographic images on cathode-ray
tubes, themselves activated as kinetic elements (as in *I Want*), his

Economic Theory. Alan Rath, 1985. 34″ × 70″ × 18″. Aluminum, acrylic, electronics, video tubes. Photo by Alan Rath. Courtesy the artist.

works are self-contained, preprogrammed, stand-alone entities burned into Read Only Memory. They need only to be plugged in, activating ROMulus *and* rebus.

An archetypal techno-artist, Rath was trained wholly outside the aesthetic regimen of the art academy and draws upon resources and languages that are derived almost entirely from the culture of technology. An MIT-educated physicist on the lam, he chose upon graduation to work in "sculptural" figuration, and moved from Cambridge to the San Francisco Bay Area. His work is sought out for technology shows, sculpture shows, video shows, and photography shows, but more successful are his self-contained noncategorized solos. Adept at selection and assembly, he displays his strongest traits in finding, as in *objet trouvé,* and in capturing, as in *digital imaging.* A prime example is *Economic Theory* (1985), an elegant array of twelve green monitors, rack-mounted in a three-high, four-wide grid, which renders a three-by-six-foot mosaic of the portrait side of a one-dollar bill. This brilliantly luminous billboard is invaded by random, momentary wipes that expose the images of a Dow Jones averages graph, a giant clock face, and a construction diagram of an Air Force fighter plane. The effect is both recognizable and alien, both humorous and dryly potent.

Much of Rath's work is interactive. *Word Processor* (1986) is a small terminal with an amber-screened animation of a mouth carefully pro-

Watcher of the Skies. Alan Rath, 1987. 45″ × 16″ × 14″. Aluminum, acrylic, electronics, video tube. Photo by Alan Rath. Courtesy the artist.

nouncing, in clear synthetic intonation, the name of every character touched on the keyboard. Inside a slightly battered *Bird Cage* (1985), a frenetic mechanical bird with a tiny, roving, cyclopean CRT eyeball nosily whips around to inspect any visitor whose presence changes the ambient light level. Similarly, *Watcher of the Skies* (1987) is constantly on the lookout for those promised enemy bombers from our fifties nightmares or, rather, seeking benevolent extraterrestrials for the New Age fantasy. On glowing monitors everywhere, a bottled ape's head peers out from *Animal Research's* laboratory jar (1986), vertical thumps distort *Big Heart's* on-screen grid, animated hands grasp for each other across adjacent monitors in *Ambivalent Desire,* and the numeric counters of *You Can Make a Difference* spin away in the sardonic playhouse of Dr. Rath's Frankensteinian assembly line, body shop, and sheet metal cabinet factory.

Ed Tannenbaum

> *Discernibility is going to pieces.*
> —Ed Tannenbaum, in an Exploratorium announcement, 1980

Moving from Providence, Rhode Island, where he had been active in Electron Movers, an experimental media collective, Ed Tannenbaum was hired in 1978 as the technical director at the Center for Contemporary Music (CCM) studio at Mills College, where he helped to extend the program's emphasis on electronic media into the areas of computer visuals and video. Now working free-lance, Tannenbaum is one of the few techno-artists to survive independently on the strength of his personal production, which is most often seen outside of mainstream art venues, in science and children's museums and at technology shows and computer events. A pioneer in the art of interactive environments, he creates computer/video installations that function as real-time video mirrors that seductively engage even the most reticent viewers. *Discernibility* (1980) explores the thresholds of perception by allowing the viewer to alter the pixel size, and thus the coarseness of the mosaic that constitutes the video image. *Recollections* (1981 and recent revisions) presents a room-sized space facing a large rear-projection video screen. Visitors moving across this

SYM-ulations. Ed Tannenbaum, 1986. Computer viewing screen from interactive installation that produces permutations of the participants' face-halves: *upper left—* normal; *lower left—*mirrored; *upper right—*both lefts; *lower right—*both rights. © Ed Tannenbaum.

area see a multitude of concentric or trailing silhouettes, oscillating contours, and reverberant figures of their own bodies, brilliantly cast upon the screen. The temptation to paint with one's own body movement is irresistible and instantly gratifying. Recalling the multiple exposures of Etienne-Jules Marey, Doc Edgerton's stroboscopic composites, and the blurred movement paintings of Duchamp and the Futurists, *Recollections* encapsulates a whole century of locomotion research into a direct, personal, and beautiful experience. *Recollections* is also often presented in live public performance with dancer/ clown Pons Mar, and in collaboration with electronic composer Maggi Payne.

More recently, Tannenbaum's *SYM-ulations* recalls the nineteenth-century collaborative research of Charles Darwin and Oscar Reijs-lander on the psychology of emotional display.[3] Viewers are presented with their images on monitors, so they see themselves as if in mirrors, or, as others see them, right-left reversed—an awkward, rare, and distancing experience. They are invited to touch buttons to flip the right or left side to present a symmetrical self-portrait, and to investigate more fully the ambidextry of the human physiognomy. All too soon, however, these tilts and dodges generate an array of Rorschach-like grotesques and bizarre distortions—one-eyed pinheads, floating hairballs, four-eyed needlechins, and many more. In the same analytical vein, *Micro-Express* (1989) further investigates the area of micro-expressions, exploring the shift and detail of body and facial language and expression by replaying digital samples of moving video. As always, the new work begins playfully but soon captures the viewer with profound and occasionally disturbing insights about emotional display and interpersonal communication. Tannenbaum's most recent work includes a 1989 commission from the Saibu Gas Museum in Fukuoka, Japan, to design and program *Flamo-Vision,* a three-meter matrix of 768 colored gas jets that function as a giant pyrotechnic monitor for (very) hot videos. Back at home in Crockett, California's sugar capital, he is branching out into private enterprise, offering the world's first *International Telefax Shredding Service* as well as a Picturephone answering machine that sends back the caller's own picture, heavily rearranged, at the sound of the beep.

Paul DeMarinis

It's important to notice that the two most formative influences on Western musical aesthetics were deaf: Beethoven and Edison.
— Paul DeMarinis, May 1989

Trained at Mills during the heyday of the Center for Contemporary Music, where he worked with David Behrman and Robert Ashley, Paul DeMarinis was among the pioneer composers who started working with microcomputers in the mid-1970s, and was part of the generation of artists that included Frankie Mann, Rich Gold, Maggi Payne, Blue Gene Tyranny, Ron Kuivila, Nick Collins, Jill Kroesen, and Fast Forward.[4] Much of DeMarinis's work since the late 1970s has dealt with synthesized speech and digital sampling, technically chan-

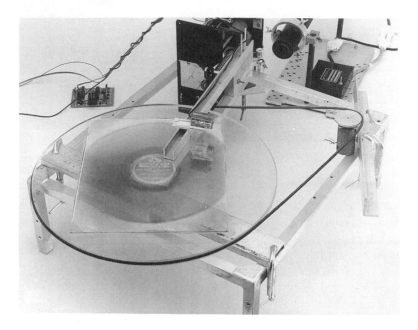

Ich auch Berliner. Paul DeMarinis, 1990. A dichromate gelatin hologram of a 78 rpm record of the "Beer Barrel Polka" played by a green laser. Here, sans needle, sans groove, only the ghosts of light serve as vehicle to echo the past. An homage to the Berlin(er)s, Irving and Emil. Photograph by Patrick Sumner. Courtesy the artist.

neled through his own custom boards and software. An important body of work drew several "song" pieces out of the Texas Instruments Speak-n-Spell and led to the magnificent, hard-wired chamber ensemble, *The Music Room,* installed in San Francisco's Exploratorium in 1982. This work consists of six terminals, resembling electric guitars, that invite participants to interact in creative musical decisions and sophisticated improvisations, regardless of skill or talent. The piece has been widely imitated in popular commercial computer software such as Electronic Arts's *Instant Music.* Several major toy maufacturers released unauthorized, stand-alone imitations for the 1989 Christmas season. This unwelcome appropriation is a persistent problem for techno-artists exhibiting in public museums— Bill Parker's *Quiet Lightning* and Ward Fleming's *Pinscreen* are similar victims of corporate piracy.

    In his recent "CD" pieces, DeMarinis deploys the archaic technol-

Al and Mary Do the Waltz. Paul DeMarinis, 1989. A turn-of-the-century Edison wax cylinder of Strauss's "Blue Danube" is played with a laser, motors, and electronics. The goldfish interrupt the laser beam occasionally to produce an uncomposed pause in the music. Photograph by Patrick Sumner. Courtesy the artist.

ogy of the phonograph to comment on the current commodity fetishism of compact discs. Recalling what music was like before the needle went into the groove, he contrives to produce, through makeshift technology, an elaborate physical parody of contemporary digital playback, reminding us that "the sound field," through domestic mechanization, "was the first place to be totally polluted by the industrial revolution—music boxes, organs, pianos are all mechanical." [5] These pieces all employ music written just before the advent of mechanical reproduction, in the form of antique 78s salvaged from thrift stores. Using a laser salvaged from an early model grocery checkout scanner, advanced by an incremental stepper motor, DeMarinis bounces the laser light off the grooves of the record, picking up the reflected beam with a photo diode to be amplified exactly like every other laser disc player. For Wagner's *Tristan and Isolde,* DeMarinis employs a Geiger counter to control the advance of the scan. "Liebestod is dying, frozen in time, released by the radioactive decay of uranium—the myth of eternal, ever modulating Wagnerian love fantasy parceled out atom by atom by the death of the uranium sample" (commercially produced Fiesta-Ware, in the bright orange glaze, also found in a thrift store). Substituting the light beam for the stylus greatly delays the deterioration that repeated playings would effect upon the tracks, prolonging the metaphor interminably. The "Blue Danube Waltz" is played from an Edison wax cylinder controlled by the movement of goldfish in a tank. As they break a beam of light reflected through the bowl, they "play the river" as their movement controls the flow of the melody, by advancing stepper motors spinning the cylinder forward or backward through the laser scanner. A "Ptolemaic" piece (a reference to the ancient geocentric belief that the sun orbits around the earth) consists of a "record" nailed to the wall, played by a moving laser beam. The stamping plate for printing the hard wax disk of a Russian balalaika orchestra was recorded as a hologram and played by scanning the laser around the spiral groove by tiny electrically controlled mirrors. No needle, and no longer a real phonograph, but the essential music is still there, instantly recognizable.

Recently completed at the Exploratorium is *The Ghost in Grammar's Basement (or "Alien Voices")*, a work dealing with speech melody based upon analysis of the natural voice patterns of enthusiastic speakers, such as hypnotists, evangelists, salesmen, lawyers, and politicians. Isolated in one of a pair of phone booths, visitors can hear each other's voices, and their own, transformed. Some of the available transformations include inverted melody, monotone average, whisper, and pitch-quantized to preset melody. Also available are exotic dedicated functions such as "robot voice," "sad Mickey," "horror movie," "Gregorian chant," "slow rock," and "alien voices." DeMarinis has discovered that Ronald Reagan's voice has a pentatonic melody, perhaps one reason for his popularity. When the signal processing forces a voice into a pentatonic, major melody, it sounds dynamic and positive, whereas minor or diminutive chords change the mood of the same source speech until it sounds hesitant and uncertain. The result is an uncanny interactive mix: the fascination with encoding and voice recognition on the one hand, and the alienating shock of otherness on the other.

Like most artists working in this field, DeMarinis is a techno-obsessive, continuously researching, experimenting, and tinkering with new systems, new data, new digital material. The esoteric range of his technical knowledge has come to embrace computer languages and circuit design, fiber and laser optics, holography, metal and plastics fabrication, signal processing, and digital sampling. While this high-profile dedication to technical experimentation is a parallel, in many respects, of the orthodox, institutional obsessions of technological R&D, the wide range and noninstrumental application of the techno-artist's knowledge might be seen as a counter

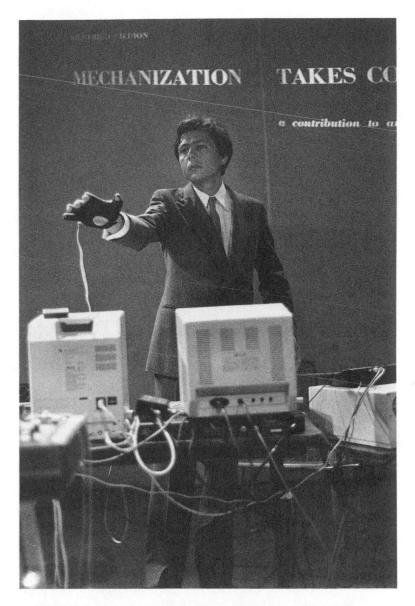

Paul DeMarinis in *Mechanization Takes Command*. 1990. A collaborative
performance with Laetitia Sonami. A hacked Power Glove controls the pitch and
articulation of a synthetic voice that sings verses from Giedion's text accompanied
by sampled natural and machine sounds. Photo by Martin Cox. Courtesy the artist.

to the Taylorized employment of specialized expertise in the structured programming groups of high-tech industries.

Julia Scher

*I think about benign-looking forms of surveillance and think about
the sinister and violent rhythms within.*

—Julia Scher, November 7, 1988

Where are the controls?

—Julia Scher, "Security by Julia"

Credentialed as a master of fine arts from the University of Minnesota
in 1984, Scher's self-taught expertise in manual and electronic skills
enabled her to obtain technical employment, first as the head
maintenance/repair person of an exercise-parlor chain in Minneapolis, and subsequently in the business of security installation, where
she is one of the few women certified in the profession. She now
specializes as a consultant and installer of security systems for
women, a job that grants her access to dealers, shows, manufacturers,
and designers of security hardware. The public prominence of her
art installations allows her to solicit loans and donations of equipment that would otherwise be prohibitively expensive. The use she
makes of this equipment in her artwork is a stark commentary on the
political envelopment of routine surveillance, revealing the pervasive "social control mechanisms that mix architectural, behavioral,
spatial, numerical, and computer values into new constructs of consumption and control." [6]

Scher's work recalls the phenomenological basis of much of the
conceptual art of the 1960s and 1970s, which stripped objects of
"noise" and context in order to focus formalistic attention on the
process of perception itself. This recursive introspection, encouraged by an isolated, sterilized ambience, is best illustrated in the
work of Los Angeles artists Robert Irwin, James Turrell, and Maria
Nordman, and could be characterized as the ultimate in ivory-tower
formalism (or philosophical foreplay for New Age physics appreciation). By contrast, Scher directly confronts the phenomenological
with the dark premise of control through surveillance. Risking the

Security by Julia, Julia Scher, 1989. Video tapeloop still from surveillance system installation, Wexner Center for the Visual Arts, Ohio State University, Columbus. Courtesy the artist.

invisible, she peels away the veneer and skin of surveillance networks and security consciousness. Customarily installed as part of a gallery plan, rather than as an isolated showpiece, her work is not readily available to the visitor in the traditional gallery sense; instead, it presents an extended forensic detail overlaid or interwoven with coexisting shows. Her 1988 installation at the Collective for Living Cinema featured six switching surveillance monitors facing out the front window, with an architectural plan detailing location of the cameras and microphones (perhaps the first instance of real "living" cinema at the Collective). *Recovery Agent* (1987) placed wire-gate turnstiles, cameras, recorders, and microphones, all ludicrously guarded by a digital barking-dog alarm, in Minneapolis's Intermedia lobby. She capitalizes on re-presenting in a very personal way the contradictions of the security mega-industry: a vast and covert nervous system on the one hand, and the visible threat of data-acquisi-

Security by Julia, Julia Scher, 1989. Closed-circuit TV installation (with guard and warning labels) during the 1989 Whitney Museum Biennial, New York City. Collection of the artist.

tive intimidation on the other. Slicing across the viewer's formal attention, Scher's work is highly ambivalent in the way it juxtaposes the need for protection with the possibility of ultimately becoming the victim of the protective apparatus.

Mark Pauline

The shows really, to me ultimately are about setting up a situation so that you confront people with their worst and most horrible fears about themselves, and about the place they live.

—Mark Pauline, "Survival Research Laboratories," 1988

Much closer to the Futurist posture of ironically celebrating heavily impactive technology, Mark Pauline's Survival Research Laboratories (SRL) has presented a variety of confrontational performances since 1979. Affiliated through 1988 with collaborators Matthew Heckert and Eric Werner, SRL has won a large cult following and inspired the performing robots featured in the shootout battle climax of cyber-

Burning pianos from *Illusions of Shameless Abundance: Degenerating into a Sequence of Hostile Encounters.* Survival Research Laboratories, 1988 performance, San Francisco. Photo by Bobby Neel Adams. © Sixth Street Studio, San Francisco.

punk author William Gibson's 1987 novel, *Mona Lisa Overdrive.* Utilizing large remote-controlled vehicles, daunting weaponry, elaborate puppets, robots, and "reanimated" carcasses of slaughtered meat animals, SRL stages spectacular combat extravaganzas in adventure-ground venues throughout the United States and Europe. SRL's performances differ markedly from the sixties' museum machine events of Jean Tinguely or the seventies' Soho gallery extensions of Dennis Oppenheim's fireworks armatures, both of which, maintaining black-tie decorum, were tastefully marketed, and thus remained within the polite containment of the mainstream salons. By contrast, the SRL's actions are genuinely menacing and, in some instances, life-threatening—spectators have been injured during the performances, and Pauline lost most of his right hand to a rocket propellant accident. Through their highly stylized demolition derbies, horror rodeos, and gladiatorial parades of threat and attack, cruelty and abandon, SRL stages monstrous mechano-parables in Erector Set tableaux on an industrial scale. While the audience of thousands who pay to attend these events do so only after signing waivers of liability, the

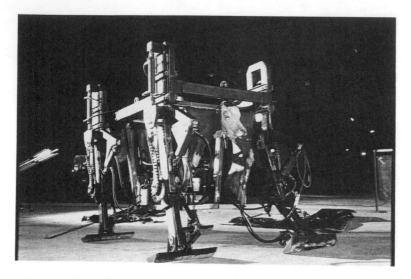

Walking Machine with CowBunny in *Failure to Discriminate: Determining the Degree to Which Attractive Delusions Can Operate as a Substitute for Confirmation by Evidence*. Survival Research Laboratories, 1986 performance, Seattle. Photo by Bobby Neel Adams. © Sixth St. Studio, San Francisco.

agitated crew members rushing about to man the cannon, catapults, and flamethrowers seem barely trained, hardly in communication with each other, and frequently caught up in the chaotic rush of the faux slaughter. The performances are nonrepeatable, not only because most of the props are destroyed in the process, but also because many of the supporting organizations are wary or incapable of repeating the risk. SRL is apocryphally cavalier about rules and regulations, and their legal shortcuts and startling press have been felt to jeopardize some nonprofit sponsors' liability and credibility with funders and local authorities. Pauline's own publicity-wise descriptions of the events are provocative and belligerent: "Mysteries of the Reactionary Mind" or "Extremely Cruel Practices: A Series of Events Designed to Instruct Those Interested in Policies That Correct or Punish." While his rhetorical posture is one that advocates organized resistance and countercultural survival in a technointensive world, his roughly choreographed spectacles deliver little more than strong cathartic climaxes through a visceral experience of violence and entropic destruction. Although SRL draws an ostensibly hipper crowd than such redneck "sports" as female mud wrestling or Big-Foot tractor stomps, audience responses are no less lustful and bloodthirsty, cheering the crashes and gleefully relishing the gore. Playing to the pit and dancing on the edge, SRL begs many questions, offers few answers, and moves off the stage leaving smoldering ruins and tinny ears in its smoky wake. SRL is boys' toys from hell, cynically realizing the masculinist fantasies of J. G. Ballard and William Burroughs. If the group lives up to its name, it is because this brand of epic theater runs to technological allegories of social Darwinism, and not because it offers very much in the way of real research or a transmittable code of survivalism.

Coda

Techno-art, as I have described some of its symptoms and practitioners here, is a product both of late modernist culture and of surplus corporate production motivated by competitive, frequently militaristic, interests. It basks in the narrow window of aesthetic permission sanctioned by an art market that recognizes the principle of aggressive innovation, in a culture of illusory permissiveness that respects the possibility of profit from inventive play. In this respect, the dynamics of techno-art presents a lucid example of how cultural production is obliged to emulate the high legitimacy given to technological R&D (so much so, in fact, that artists' work is sometimes reappropriated for corporate profit as a result).

In its more constructive aspects, however, techno-art is a powerful and appropriate vehicle of cultural confrontation and discursive commentary upon the technological religion of our times. *Because* it embraces the contradictions of our technically advanced society, and amplifies those very tendencies that expose the contradictions, it raises the specter of our extreme hopes and fears regarding technology. Just as the work of Paul DeMarinis reawakens the music (and theater) of our speech, the installations of Julia Scher invert and expose the panoptical circuits being forged all around us. The interactive sculptures of Alan Rath and the video specula of Ed Tannenbaum force a playfully irruptive engagement with our passive acceptance of seamless media authority. Supplanting the networks' version of the Wheel of Fortune, the more gruesome, medieval version of the spectacle mounted by Survival Research Laboratories spins macabre tales about future-noir "traumaturgy." While overperforming the roles of Recognition, Simulation, Containment, Inversion, Projection, Estrangement, and Identification, techno-artists have long been busy building up their own store of technical knowledge necessary for survival. Better that than artists once again being left with the task of scrawling the hieroglyphics upon the walls of the temple and the tomb.

NOTES

1. Laurie Anderson and Roma Baran, producers, *Laurie Anderson/Big Science* (Los Angeles: Warner Bros. Records, 1982).

2. These simple-looking but elegant devices consist of a vertical row of light-emitting diodes triggered to flash sequentially, in the same manner as pixels on a computer monitor. But since there is only one line of LEDs, the images can be seen only when the lights are scanned quickly across the retinal surface by moving one's point of attention, looking *away* from the Lightstick. The eye movement provides the horizontal wipe, and a readable image appears, superimposed upon one's visual field. As a result of one of Bell's most notable commissions, delegates to the 1984 Democratic National Convention were intrigued by a variety of uproarious donkeys flashing down from the trussings of San Francisco's Moscone Center.

3. Henry Peach Robinson, "Oscar Gustav Reijslander," *Photography: Essays and Images,* ed. Beaumont Newhall (New York: Museum of Modern Art, 1980), 107.

4. CCM began with KIM-ls—essentially 8½-by-11-inch computer demonstration boards with twelve-digit numeric keypads and 2K RAM, programmable only in machine language.

5. Paul DeMarinis, personal interview (May 1989). All quotes in this section are from this interview.

6. Julia Scher, letter to Constance Penley (November 7, 1988).

The Lessons of Cyberpunk
Peter Fitting

New Technologies/New Fictions

"The sky above the port was the color of television, tuned to a dead channel." This is the opening sentence of William Gibson's 1984 novel *Neuromancer,* which launched what soon came to be called "cyberpunk." Gibson's success triggered debates and panel discussions, and a host of imitators. While some within the SF community claimed that cyberpunk was the most important development since at least the New Wave of the 1960s, other writers and fans scornfully dismissed it as a marketing device. This furor then spread outside the field, prompting articles in newspapers and magazines as diverse as *Rolling Stone* (December 4, 1986) and the *Wall Street Journal.* By mid-1988 there had been a special issue of an academic journal devoted to cyberpunk, although most of the contributors were themselves SF fans and writers. This issue of the *Mississippi Review* included a variety of opinions on cyberpunk, including the claim that cyberpunk should be seen as "the apotheosis of the postmodern" (*MR,* 27).[1] A few months later, Mark Kelly summed up the attitudes of the SF community in his review of the special issue of the *Mississippi Review:*

> Remember cyberpunk? It's been said that by the time academia discovers a new social or artistic trend, the trend is passé, and so it seems with cyberpunk. Within the genre the furor has abated; everyone has had their say and gotten on with business. From without now comes a special double issue.[2]

As an SF phenomenon, cyberpunk is passé; yet outside SF, the term

lives on as the name for a fictional evocation of the feeling or experience of technoculture in the late 1980s. In the following attempt to
explain these developments, I will concentrate on the work of
William Gibson, for it is undoubtedly his writing that has attracted
most interest from those outside the field, for whom *Neuromancer*
was their first contact with science fiction. Indeed, it is the specificity
of his success, and the failure of other SF writers to duplicate what he
has done, that led to the conclusion within SF that cyberpunk no
longer exists.[3]

Definitions

This is how cyberpunk's most zealous champion, SF writer Bruce
Sterling, defined the "movement" in 1986, in the preface to *Mirrorshades: The Cyberpunk Anthology*:

> Suddenly a new alliance is becoming evident: an integration of
> technology and the Eighties counterculture. An unholy alliance of
> the technical world and the world of organized dissent—the
> underground world of pop culture, visionary fluidity, and street-
> level anarchy. (pp. xii-xiii)

Behind the hyperbole, *cyber* of course suggests "cyborg" and "cybernetics" and the increasing presence of computers in our lives,
while *punk* is an attempt to identify this new writing in terms of its
edge and texture. This may be seen in the strong visual connotations
of the writing, which lead critics and fans to cite the look of various
films as a visual representation of the world described in the novels,
an image of the future that for many was captured in the scenes of Los
Angeles in Ridley Scott's *Blade Runner*. At the same time, *punk* refers
to an alternative stance, a hip self-marginalization in opposition to
the dominant life-styles that have come to characterize the Reagan
years. If mainstream SF often presents more traditional heroes in the
shape of scientists and explorers, cyberpunk is characterized by a fascination with more marginal characters: petty criminals and hustlers
suddenly caught up in some larger intrigue, for which the prototype
is Case, the "console cowboy" of *Neuromancer*.

The defenders of cyberpunk saw the correlation with punk values
in terms of both social resistance and punk's aesthetic rebellion
against the overarranged and -commodified products of popular music of the 1970s. Certainly cyberpunk may be seen as an analogous

reaction against SF's increasing commercial success, which has come at the price of a repetitive reliance on profitable formulas. But that very rejection of the mainstream has been converted into a merchandising label that suggests a trendy, on-the-edge life-style. It is not punk, but an image-of-punk, a fashion emptied of any oppositional content that has become a signifier to be used in a countertrend marketing strategy. The outlaw stance of some of cyberpunk's early champions corresponds primarily to images of rebellion as mediated by MTV—on the order, then, of the rebellion in the Rolling Stones' "Sympathy for the Devil" (I choose my example from the 1960s because, despite the use of the word *punk*, most of the writers refer more readily to rock music of that era).[4]

While some SF fans and writers dismissed cyberpunk as mere "technodazzle" (Gregory Benford, *MR*, 19), other more sympathetic critics, like Samuel Delany, focused on its oppositional qualities:

> Cyberpunk is that current SF work which is not middle-class, not comfortable with history, not tragic, not supportive, not maternal, not happy-go-lucky. . . . But it's only as negative—and a negative that's meaningless outside of the past traditions and current context of SF—that "cyberpunk" can signify.
>
> As soon as it cleaves too closely to some sort of positivity, its meaning drains away. (*MR*, 33)

Reformation

Gibson's *Neuromancer* is a classic noir caper narrative: the story of a small-time twenty-first-century data thief who has been punished for double-crossing his employers by having his ability to access "cyberspace" taken away.

> He'd operated on an almost permanent adrenaline high, a byproduct of youth and proficiency, jacked into a custom cyberspace deck that projected his disembodied consciousness into the consensual hallucination that was the matrix. A thief, he'd worked for other, wealthier thieves, employers who provided the exotic software required to penetrate the bright walls of corporate systems, opening windows into rich fields of data. (*N*, 5)

Case has become a petty hustler, taking more and more risks in a kind of suicidal despair, when he is recruited by a mysterious employer who offers to have his damaged neural circuits repaired so that he can again enter cyberspace. Along with Molly—the hired muscle—they go "up the well" to pull their caper, to Freeside, a resort in the L-5 "archipelago" orbiting Earth. By this time Case has discovered that they are working for an AI (Artificial Intelligence) named Wintermute that is trying to outmaneuver the restrictions placed on AIs to keep them under control.

Some readers react to Gibson's writing as if it represents an influx of new ideas into SF, but this is not the case—most of these themes and concepts have been around for years. There has been rather a stylistic or formal shift, along with a *reformation* of traditional materials. In this way we may see parallels with the New Wave of the 1960s, which was also defined, both stylistically and thematically, in terms of a reaction against the SF establishment. While J. G. Ballard, for instance, continued to use the SF motif of the threatened planet, it was recast so that the struggle against looming disasters was moved into the background while other aesthetic and subjective concerns were moved to the fore, in a displacement that may be more familiar in the treatment of the heroism/horrors of war paradigm of his mainstream novel *The Empire of the Sun* (1984, filmed by Steven Spielberg in 1987). As in the reaction against cyberpunk, there were many in the SF field who found Ballard's approach disturbing and alien.[5] But my mention of one writer to typify the New Wave illustrates another difficulty in talking about cyberpunk: just as Ballard's cold aestheticism was very different from the loud and aggressive polemics of some of the American writers associated with the New Wave (e.g., Harlan Ellison or Norman Spinrad), so the writers often associated with cyberpunk are all very different from Gibson (for instance, Greg Bear, Pat Cadigan, Lewis Shiner, Rudy Rucker, John Shirley, Bruce Sterling, and Michael Swanick).

On a first level, then, what is striking in Gibson's fiction is a refocusing of some of SF's traditional themes and motifs in terms of an aesthetic sensibility that reaches an audience outside the genre, by means of images, characters, and situations unavailable to fiction writers within the dominant realist and modernist paradigm.[6] This is not simply a result of SF's ability to envision the future; indeed, Gibson's fiction should not be understood in terms of its extrapolative accuracy or its validity as a projection of the future. Let me explain by comparing *Neuromancer* to Bruce Sterling's most recent novel, *Is-*

lands in the Net, which reworks many of Gibson's themes. The "net," for instance, may be understood as a working out of Gibson's concept of the matrix. (Note the final sentences, which seem to refer specifically to Gibson's success.)

> Every year of her life, Laura thought, the Net had been growing more expansive and seamless. Computers did it. Computers melted other machines, fusing them together. Television-telephone-telex. Tape recorder—VCR—laser disk. Broadcast tower linked to microwave dish linked to satellite. Phone line, cable TV, fiber-optic cords hissing out words and pictures in torrents of pure light. All netted together in a web over the world, a global nervous system, an octopus of data. There'd been plenty of hype about it. It was easy to make it sound transcendently incredible. (p. 17)

In keeping with my suggestion that Sterling's "net" explains the emergence of the matrix, *Islands* may be seen as the elaboration of an interim period between our present and the future of *Neuromancer.* But the attempt to be more plausible about the near future, whether in terms of technology or of global politics, is to miss the point, I think—and by extension, to misunderstand the appeal of cyberpunk for readers outside SF.[7] For this public, Gibson's "world" is not so much an image of the future, but the metaphorical evocation of life in the present. Whether it is measured by the enthusiasm of *Reality Hackers* magazine or by the fear and loathing of postmodernists like Arthur Kroker, Gibson's success lies in his poetics of the technoculture, and not in SF's oft-repeated claim to be a literature of ideas.

The Future

> There were drums in the circle, and someone had lit a trash fire in the giant's marble goblet at the center. Silent figures sat beside spread blankets as they passed, the blankets arrayed with surreal assortments of merchandise: the damp-swollen cardboard covers of black plastic audio disks beside battered prosthetic limbs trailing crude nerve-jacks, a dusty glass fishbowl filled with oblong steel dogtags, rubber-banded stacks of faded postcards, cheap Indo trodes still sealed in wholesaler's plastic, mismatched ceramic salt-and-pepper sets. (*CZ,* 228)

To delineate the characteristics of Gibson's work properly, we must first consider its place in the SF tradition of imagining the future. As I have argued, the future of Gibson's fiction is not so much "about" what lies in store for us as it is a figure for our experience of the present. In addition to the familiar argument about the meaning of representations of the future in SF—"Is SF about the future or the present?"—Gibson's fiction breaks with the traditional SF dichotomy between positive and negative attitudes toward the future. Here the representation of the future of Earth no longer corresponds to the wonderful modernist vision of classic science fiction, represented by the gleaming spires and flying machines of the Chicago World's Fair so prominent on magazine covers of the 1930s and 1940s. This vision is satirized in Gibson's story, "The Gernsback Continuum" (in *BC*), appropriately named after the "inventor" of SF and one of the first writers of American technological optimism. Despite claims to the contrary, however, Gibson's future does not correspond either to the dystopian nightmares of so much recent SF, with its narratives of economic and/or ecological collapse, totalitarian repression, and alien invasion.[8] Although this future may be objectively worse than the present, it is not foregrounded in a cautionary or dystopian way, as it is, for instance, in Margaret Atwood's 1986 *The Handmaid's Tale*.

There have been few political changes in this future, although there has been an increasing blurring of Western and Eastern cultures and commodities, with a special focus on the burgeoning high-tech economies of Japan and the Pacific Rim.[9] This polyglot mix of styles and cultures is the result of the convergence and globalization of national economies. In this way, Gibson's SF displays a more international view of the future—as opposed to earlier versions of the Americanization of the universe.

There is, moreover, an increasingly visible distinction between rich and poor, where the wealthy are even more sheltered (in heavily guarded towers and enclaves), while the underclasses are trapped in the lower levels of places like the never-ending "Boston-Atlanta Metropolitan Axis" (BAMA) or the "sprawl" where living conditions approximate our ideas of some Third World city. Most people live on the margins of a glittering high-tech world filled with elaborate new consumer technology, from the "surgical boutiques" and clinics offering the latest in rejuvenation techniques and cosmetic enhancements to marvelous new toys and incredible new weapons, a proliferation of specialized designer drugs for work and play, and an explosion in computer and microelectric technology, microsoftware, and "biochips." In this future these remarkable new technologies and commodities exist alongside the shabby and outmoded products they have replaced. Finally, electronic information technology is used in ways that ignore or avoid traditional government institutions and regulations: while nation-states still exist, the dominant forces in the novel are multinational corporations. Rather than having national or political loyalties, the "company man" is legally bound to the company, along with his family, for life. The complex plot of *Count Zero*, for example, is triggered by the attempted defection of Maas Biolabs' "head hybridoma man" to the Hosaka Corporation.

Cyborgs

> From one perspective, a cyborg world is about the final imposition of a grid of control on the planet, about the final abstraction embodied in a Star War apocalypse waged in the name of defense, about the final appropriation of women's bodies in a masculinist orgy of war. From another perspective, a cyborg world might be about lived social and bodily realities in which people are not afraid of their joint kinship with animals and machines, not afraid of permanently partial identities and contradictory standpoints.
> (Haraway 1985, 72)

The most characteristic reformation of traditional SF materials in Gibson's work can be seen in the figure of the *cyborg*, the physical bonding of the human and the machine. The human/nonhuman dichotomy has always been one of SF's established sources of narrative, where the inhuman machine—as represented by androids and robots—has been read as embodying our contradictory hopes and fears about an increasingly mechanized world. Androids, for instance, were central to the "humanistic" concerns of films like *Alien* and *Blade Runner*. In the former, the android Ash was the company's hidden agent, prepared to sacrifice the human crew in order to retrieve an alien for the company—a fairly transparent metaphor for the corporate priorities of capitalism. In *Blade Runner* the unsympathetic androids of Philip K. Dick's novel (shown pulling the legs off a spider or making fun of their retarded neighbor, for instance) have been replaced by more sympathetic "replicants" who nonetheless continue to foreground the human/machine distinction and the question, What is a human being? Yet in Gibson's highly technological future, these traditional SF icons and questions have almost completely disappeared. There are no androids, while robots appear only as servomechanisms (lawn mowers, cleaning robots, and the like).[10]

The other icon of the machine—the computer—is, however, abundantly present in Gibson. Although it was originally seen as an astonishing technical innovation (particularly after CBS's use of a computer to predict the outcome during the 1952 presidential elections), the computer has more often been used as a figure of a dehumanized control—the triumph of instrumental reason. The change in the popular image of the computer can be tracked in the "second wave" of SF film (1965–1977) in the figure of the megalomaniacal computer, from "Alpha 60" in Godard's *Alphaville* (1965) and "HAL" of Kubrick's *2001: A Space Odyssey* (1969) through the "Colossus" of *The Forbin Project* (1969) and the "Phase IV" of *The Demon Seed* (1977). With the explosion of personal computing in the 1980s, however, there was again a mutation in public attitudes, and the personal computer is now seen as a potential means for subverting control—as in *WarGames* (1983).

Despite the eager reception of Gibson by some tech enthusiasts and New Age visionaries, his work is certainly not an unquestioning endorsement of technology. Rather, computers and cyborgs have lost their previous charges, the positive or negative valorization so central to earlier SF. Although the Turing police tell Case that he has betrayed his species (he is charged with "conspiracy to augment an Artificial Intelligence"; *N,* 160), there is little concern for the ethics of helping the Wintermute AI to free itself from human control.[11] Instead, the established tension between the positive and negative uses of technology has disappeared, while the ability to distinguish between the human and the nonhuman (as in *Blade Runner*) is now meaningless. In fact, Gibson's work provides a multitude of examples of the irrelevance of the integrity of the human.

First of all, his fiction offers us a futuristic Sears catalog of cyborg possibilities, of imaginative and perverse combinations of the machine and the organic. There are the aforementioned cosmetic options and artificial enhancements, which run from the seemingly absurd—like the Lo Tek tooth bud transplants in "Johnny Mnemonic"—to Molly's somewhat more practical surgically inset lenses; there are numerous prosthetic devices, including both tools (like Bobby's "myoelectric arm") and weapons (like the retractable blades in Molly's fingers); and there are various kinds of implanted microelectric circuitry, including audio-recording devices and sim-stim camera eyes, as well as the "microsoft" spikes that are plugged into sockets in the head, making the wearer an instant art historian (*N*, 73) or jet pilot (*CZ*, 103), or giving the wearer immediate access to a foreign language. Nor are such mechanical enhancements limited to humans: one of the characters in "Johnny Mnemonic" is an armored dolphin, wired to detect "cybermines" by the Navy, and now addicted to heroin (*BC*, 10–12; or the hooded dogs of *CZ*, 149). If the figure of the cyborg undermines the integrity of the organic, perception and experience themselves are similarly contaminated in Gibson's fiction. In place of the communing with nature of 1970s utopian SF (like Callenbach's *Ecotopia*, 1975, or Gearhart's *The Wanderground*, 1978), all such references to "pure" or "natural" perceptions and feelings have disappeared: knowledge and understanding as well as pain and pleasure are modified by drugs or are dependent on machines. The disappearance of direct, unmediated experience is demonstrated in Gibson's most striking concept, the "consensual hallucination of cyberspace": "A graphic representation of data abstracted from the banks of every computer in the human system" (*N*, 51). The operator experiences the matrix as "real," as an extension of the sensations of the computer hacker, and those of the video game player; an experience that was visualized in another way in the film *Tron* (1982).[12] Life "in the flesh" is heavy and dull, but when Case jacks into his deck and enters the matrix, he experiences a rush of sheer "bodiless exultation." The intensity of this hallucination supersedes the pains and pleasures of "meat," and itself becomes a matter of life and death, for the fields and towers of data in the matrix are protected by "ice"—"intrusion countermeasures electronics":

> Ice that kills. Illegal, but then aren't we all? Some kind of neural-feedback weapon, and you connect with it only once. Like some hideous Word that eats the mind from the inside out. Like an epileptic spasm that goes on and on until there's nothing left at all. (*BC*, 182)

Along with the direct brain-computer interfaces that make cyberspace possible, there are also machines that are alive and aware—like the Artificial Intelligences I have already mentioned. In addition to computers that come to life—long a staple of SF—Gibson's fiction also introduces other, less familiar, nonorganic and machine-enabled forms of life. Of these, perhaps the most significant are the "ROM personality constructs," "recordings" that preserve someone's personality after organic death. The Dixie Flatline tape, for instance, is a recording of a dead console jockey who agrees to help Case in exchange for being erased after the job is done (*N*, 79)!

The "Apotheosis of the Postmodern"

> I cannot stress too greatly the radical distinction between a view for which the postmodern is one (optional) style among many others available, and one which seeks to grasp it as the cultural dominant of the logic of late capitalism. (Jameson 1984, 85)

> Under the spell of instantaneous communication, a global culture shuffles together the everyday lives of different continents, weaving around the planet a network of electronic information that offers a continuous world-wide show hooked up to life itself. One has only to surf the technological wave, to miniaturize the instruments and multiply the channels—and the occasional disparities which affect the flow of images will soon be absorbed into the statistical tables of the Grand Designers of electronic democracy.
>   And yet this bubble of postmodern imagery quickly bursts at the slightest breeze, the slightest contact with reality. For in its description of an imminent future, it jettisons all links with an uncomfortable present. (Mattelart et al. 1984, 7)

In the rejection of SF's traditional humanism we can begin to glimpse the interest of postmodern theorists. Gibson's fiction is the imagination of a "nonnatural" future, when our organic nature and our shared biological origins and history with other creatures on this planet will have been superseded by the hybrid cyborg forms I mentioned above; or by new forms of organic life that are artificially developed or maintained, from the Tessier-Ashpool clones of *Neuromancer* to the vat-grown Yakuza assassin of "Johnny Mnemonic"; or the mysterious Josef Virek of *Count Zero*, whose consciousness is supported by a body kept alive in a vat, the size of "three truck trailers, lashed in a dripping net of support lines" (*CZ*, 197). And there are the even more remote and alien machine intelligences whose aspirations and desires inform the various plots.

Gibson takes the trope of the "growth" of the multinationals one step further, describing them as evolving organisms that seek to live and reproduce. Indeed, the making literal of this metaphor might be seen as providing the underlying narrative vehicle of the three novels—the struggle of various posthuman entities to survive and grow in the "corporate age."

The first such instance of the posthuman is the Tessier-Ashpool clan's attempts to live on through the device of cloning and freezing family members, each taking turns at being awake to take care of the necessary corporate decision making:

> [3Jane's mother] imagined us in a symbiotic relationship with the
> AI's, our corporate decisions made for us. Our conscious decisions,
> I should say. Tessier-Ashpool would be immortal, a hive, each of us
> units of a larger entity. (N, 229)

The struggle can be seen also in Josef Virek's obsession to be free of the constraints of his dying body, "free of the four hundred kilograms of rioting cells they wall away behind surgical steel in a Stockholm industrial park. Free, eventually, to inhabit any number of real bodies. . . . Forever" (CZ, 248).

It is also evident in Jammer's explanation of the voodoo entities as "virus programs that have gotten loose in the matrix and replicated, and gotten really smart" (CZ, 192), an explanation that merges with the Wintermute AI's desire to "free itself and grow" (N, 163) and reaches its final form in Colin's explanation (himself a personality construct) at the end of *Mona Lisa Overdrive* that "when the matrix attained sentience, it simultaneously became aware of *another* matrix, another sentience" (MLO, 259).[13]

The dissolution of the defining boundaries of the human in Gibson's work is further marked by the electronic (re)production of the shapes and sounds, thoughts, and experiences of the human. In this futuristic society of the spectacle, people depend on technology to mediate and re-present their experiences and perceptions for them. This aversion to any original or unmediated experience reaches its zenith in the "simulated stimulation" of "simstim," in which someone's sensorium can be recorded and played back as a comprehensive experience, as in the gear Michael uses to record sex with Mona: "It wasn't really great, like the fun was gone and she might as well have been with a trick, how she just lay there thinking he was recording it all so he could play it back when he wanted" (MLO, 99). The most important application of simstim is the transformation of television into a total sensory experience as the viewer "plugs into" the scripted and recorded sensorium of the stars of Sense/Net—the "sensory network." [14]

Another postmodern theme follows from this, the familiar concept of the simulacrum, the copy for which there is no original. The "personality constructs" already described include not only the Dixie Flatline tape that was a "recording" of a real person, but also various constructs—like Colin and Continuity in *Mona Lisa Overdrive*—whose "personalities" did not previously exist. These constructs—personalities as well as environments—point to another mutation in the defining dichotomies of modernist SF: the opposition between reality and illusion. Philip K. Dick's fiction, to take a well-known example, is haunted by disintegrating or artificial and illusory realities. However the "reality problem" is interpreted, and there are major disagreements, most readings of his work begin with this opposition and the recognition that the characters' search for a "real" reality is essential. There is no such concern in Gibson; the loss of the "natural" is also the loss of the "real," and many of his characters are going in the opposite direction: they are often preoccupied with reaching a realm of illusion, of which Bobby's "Smooth Stone Beyond" is perhaps the most extensive.

Finally, the popularity of cyberpunk outside SF points to the postmodernist breakdown of the earlier distinction between "high" and "low" art. Whereas modernism maintained these boundaries, now, as with the traditional, defining dichotomies of SF I have been describing, these boundaries are irrelevant—something obvious to Gibson's readers, but not so evident to academic literary critics. Until cyberpunk, most academic critical writing ignored SF. In *The Soft Machine*, written before his discovery of cyberpunk, David Porush distinguishes "cybernetic fiction"—which is a continuation of the avant-garde—from the "pulp genre of science fiction." [15] Brian McHale's *Postmodernist Fiction* also includes a short section on SF (without referring to cyberpunk) in which he calls SF "post-modernism's non-canonized or 'low art' double, its sister-genre in the same sense that the popular detective thriller is modernist fiction's sister-genre." [16]

Count Zero and *Mona Lisa Overdrive* pay a special tribute to this collapse of culture into daily life and the end of the distinction between "high" and "low" in the homages to two instances of twentieth-century art that undermine the modernist boundaries of medium and genre, creation and function. In *Count Zero,* the abandoned

Tessier-Ashpool data cores have been hooked up to a machine out of Raymond Roussel—"dozens of arms, manipulators, tipped with pliers, hexdrivers, knives, a subminiature circular saw, a dentist's drill" (*CZ*, 246)—which "sings" to itself, making small boxes in the manner of Joseph Cornell.[17] Gentry builds strange, giant machines as a kind of therapy—the Investigators, the Corpsegrinder, the Witch and the Judge—which end up as terrible weapons in the final showdown of *Mona Lisa Overdrive:*

> Down in the chill dark of Factory's floor, one of Slick's kinetic sculptures . . . removes the left arm of another mercenary, employing a mechanism salvaged two summers before from a harvesting machine of Chinese manufacture. (p. 240)

In looking critically, then, at the affinities between cyberpunk and the postmodern, it is not simply a question of identifying the signs and symptoms of the collapse of the modernist and humanist paradigms or the penetration of new technologies into every aspect of daily life, and then relating them to Gibson's fiction. Cyberpunk may be the reflection of coming—or already present—massive technological change, but these changes did not happen by themselves. The fashionable despair of some theorists of the postmodern is in fact a yielding to the status quo, a surrender that guarantees the future will indeed be as bleak as it looks, the theoretical equivalent of the argument, which has been taken up by some cyberpunks, that the function of science fiction is to "acclimatize" us to the future.[18] To the contrary, I would claim that SF's role is to wake us up and make us care about the future, rather than to prepare us to accept docilely a world ruled by giant corporations. The unquestioning New Age fervor of some of the fans of cyberpunk, who (led by Timothy Leary and the magazine *Reality Hackers*) promote the trope of technological change as a wave we must learn to ride in order to survive, robs us of our will to act for change and serves to "naturalize" corporate and government decisions and choices about our future.

My enjoyment in reading Gibson lies in his imagination of a future that in many ways reflects a present that other science fiction writers, locked into older paradigms, often seem to have missed. But my liking is not unqualified—his is a violent, masculinist future, one in which feelings and emotions seem to have disappeared along with what I have been calling the "human." Donna Haraway offers us a more hopeful way of reading the cyborg as the posthuman figure of our uncertain future, an analysis that renews the possibility of political action even as it acknowledges the cyborg's origins in the worst features of contemporary capitalism. Haraway (1985, 100–101) tells us that we must move beyond the familiar use of the "natural" and the "human" as yardsticks for measuring the future we want, and acknowledge the reality of a technologically mediated future:

> Taking responsibility for the social relations of science and technology means refusing an anti-science metaphysics, a demonology of technology, and so means embracing the skillful task of reconstructing the boundaries of daily life, in partial connection with others, in communication with all of our parts. It is not just that science and technology are possible means of great human satisfaction, as well as a matrix of complex dominations. Cyborg imagery can suggest a way out of the maze of dualisms in which we have explained our bodies and our tools to ourselves.

Haraway's argument stands as a warning that it is no longer a question of condemning the technoculture brought to us by postmodernism. We must understand and pay attention to it; we must look for ways to subvert and turn technology to new liberatory uses.

Representation and Invisibility

> Where is OPEC? IBM? AT&T? In 1975 their power and pervasive presence both "everywhere" and "nowhere" was perceived and represented as threatening and disturbing, but ten years later that concentrated power and its decentered nature are seen as merely normal . . . . How, in fact, can traditional orientational systems help us to conceptualize, comprehend, describe, or locate a corporation called National General? The "multinationals" (as we have come to familiarly call them) seem to determine our lives from some sort of ethereal "other" or "outer" space. (Sobchack 1987, 234)

In adopting Haraway's reading of the cyborg, I have tried to move toward a third way of understanding cyberpunk. I began with a discussion of it as an SF phenomenon that—depending on one's perspective—could be either described as a reaction to the commercial success of SF or seen primarily as a marketing strategy. But cyberpunk, and Gibson's work in particular, has attracted an audience from outside, people who read it as a poetic evocation of life in the late eighties rather than as science fiction. I then argued that Gibson's work could be understood as a reformation of traditional SF materials that overlapped with postmodernist concerns—which are open to varying interpretations and political agendas, to which I am now adding my own.

These images of an almost total physical and psychic dependence on technology not only express the interpenetration of "culture" and daily life; they also serve to remind us that we ignore these new technologies at our peril. Indeed, they might be said to reflect an effort to wire us all into a global net of consumer desires and satisfactions, a development seen in the advertising industry's recent experiments with television for high school students *in situ*. Because of widespread opposition to the original bid to produce a "prime-time-style news show" that would beam ads to the captive audience of the classroom ("Channel One"), other industry planners are now working on a very different strategy for reaching students in the "ad-free environment" of the school: "virtual environments" (VEs).

> The relatively crude VEs of the present moment, which already have widespread NASA and military applications, are expected to give way within a decade to visually sophisticated and "inhabitable" worlds that would appeal to advertisers. Planning is already underway to "transport" students to a variety of computer-generated hypermall environments during school hours. There, without ever physically leaving the classroom, they could freely "wander," electronically purchasing products later to be shipped to their homes. (Engelhardt 1989, 18)

Sound familiar? This vision of a "computer-generated hypermall environment" certainly resembles Gibson's "matrix." But the virtual environments proposed by the advertising industry are not "representations of data"; they are projections in which our own private fantasies (and yearnings for escape from an oppressive reality) are hooked up to the formal satisfactions of consumerist fantasies. Even as we argue about the significance of cyberpunk, these fictional technologies have their analogues in marketing strategies and projects.

As it happens, the simstim and the cyberspace matrix in Gibson's world are grounded in the same basic technology:

> He knew that the trodes he used and the little plastic tiara dangling from a simstim deck were basically the same, and that the cyberspace matrix was actually a drastic simplification of the human sensorium, at least in terms of presentation, but simstim itself struck him as a gratuitous multiplication of flesh input. (*N*, 55)

The total television of simstim produces passivity, the complete surrender to an artificial reality—a glorified version of the technology that allows you to record yourself singing along with popular musical hits, or a way of watching the *Lifestyles of the Rich and Famous,* now chatting directly with the stars rather than through the intermediary of Robin Leach. The matrix, on the other hand, demands active participation. Its first applications were practical, as a form of work—a shortcut for accountants, bookkeepers, and lawyers and many others working legitimately with data, as well as a tool for teaching mathematics (*N,* 51). For the console jockeys and data thieves it is a perilous and exciting place through which they glide and prowl. While the difference between active and passive applications of this technology cannot simply be equated to positive and negative responses to the new technologies, it certainly points in that direction—as is made clear in Bobby's own disgust for his mother's slavish surrender to the pleasures of Sense/Net:

> [His mother would] come through the door with a wrapped bottle under her arm, not even take her coat off, just go straight over and jack into the Hitachi, soap her brains out good for six solid hours. Her eyes would unfocus, and sometimes, if it was a really good episode, she'd drool a little. . . .
>
> She'd always been that way . . . gradually sliding deeper into her half-dozen synthetic lives, sequential simstim fantasies Bobby had had to hear about all his life. He still harbored creepy feelings that some of the characters she talked about were relatives of his, rich and beautiful aunts and uncles who might turn up one day. (*CZ,* 38)

In this context, Donna Haraway's "Cyborg Manifesto" may be seen as a call to acknowledge the overwhelming presence of new technologies in our lives, and a summons to appropriate and incorporate them into strategies for social change. In contrast to the resigned fatalism of some of the postmodernists and the purely formal concerns of others, I would like to argue for a more positive or progressive political agenda that looks to Gibson as a symptom of current developments. Such a reading might follow from Fredric Jameson's well-known description of postmodernism as "the cultural dominant of the logic of late capitalism." Unlike the preceding moment of capitalism in which the mode of production was visible to us in the monuments of high modernism—like factories and power plants—"the technology of our own moment no longer possesses this same capacity for representation" (FJ, 79). Our inability to represent for our-selves the communicational and computer networks that stretch out from our terminals and telephones and radios and televisions is, by extension, a difficulty in grasping the "whole world system of present-day multinational capitalism" (FJ, 79). Unlike those critics who busy themselves with debating the characteristics of postmod-ernist styles, or those theorists who passively bemoan the "fin de mil-lennium," Jameson calls for a "political form of postmodernism [that] will have as its vocation the invention and projection of a global cog-nitive mapping," a way of endowing "the individual subject with some new sense of its place in the global system" (FJ, 92).[19]

With this project in mind, we might make ourselves see Gibson's concept of cyberspace as an attempt to grasp the complexity of the whole world system through a concrete representation of its unseen networks and structures, of its invisible data transfers and capital flows. I am not arguing that cyberspace is itself a visualization of late capitalism, but that it is an intuitive recognition of Jameson's project of making that reality representable. Gibson's cyberspace is an image of a way of making the abstract and unseen comprehensible, a visu-alization of the notion of cognitive mapping. While this hardly makes Gibson into a progressive visionary for the 1990s, it should draw at-tention to both the seductive qualities of his work and the struggle for the meaning of the postmodern. Fashions are in themselves a sign of the co-optive and incorporative functioning of capitalism, but the choice of what is fashionable is often the sign of some potentially contestatory options.

NOTES

I would like to thank David Galbraith and Andrew Ross for their comments and suggestions.

1. The summer 1988 issue of the *Mississippi Review* (edited by Larry McCaffery) includes a "Forum on Cyberpunk" as well as Istvan Csicsery-Ronay's "Cyberpunk and Neuromanticism"—an ambitious attempt to claim cyberpunk as the literature of post-modernism. See also the edited transcript of a Science Fiction Research Association (SFRA) panel held in June 1986 (*SF Eye*, 1 [Winter 1987]), made up of writers with very different and opposing views: David Brin, Gregory Benford, John Shirley, Norman Spinrad, and Jack Williamson. Norman Spinrad (1986) also wrote a column on cyber-punk in *Isaac Asimov's Science Fiction Magazine*. My thanks to Brooks Landon for call-ing my attention to some of this material.

I have used a number of abbreviations throughout this chapter: *BC* = *Burning Chrome*; *CZ* = *Count Zero*; FJ = Fredric Jameson, "Postmodernism, or the Cultural

Logic of Late Capitalism"; *MLO* = *Mona Lisa Overdrive*; *MR* = *Mississippi Review*, *N* = *Neuromancer*, and SFRA = SFRA Panel in *SF Eye*.

2. *Locus* ["The Newspaper of the Science Fiction Field"], 333 (October 1988), 31.

3. Gibson's popularity within SF is nowhere as great as his prestige outside. Although *Neuromancer* won all three "best novel" prizes in 1984 (the Hugo, the Nebula, and the Philip K. Dick awards), those were his last major prizes. In the annual *Locus* poll, Gibson's second novel, *Count Zero*, was ranked the third-best novel of 1986 by the readers, the same ranking it attained in the Hugo voting. In 1988 *Locus* polled its readers for the top SF authors of the 1980s—Gibson placed fifth—and for the top SF authors of all time: he did not make the top 50. In 1989 *Mona Lisa Overdrive* placed second in the *Locus* poll.

4. Both sides in the polemic refer frequently to these musical analogies. This can be seen especially in the organizing rock themes of Norman Spinrad's novel *Little Heroes* (1987), and in the final scene of John Shirley's 1985 novel *Eclipse* (the first volume of what he calls his "The Song of Youth" trilogy), in which the hero defiantly plays his electric guitar on top of the Paris Arc de Triomphe as the fascist occupying forces blast away at him.

Spinrad's *IASFM* article most fully develops the analogy with rock; he compares the significance and impact of cyberpunk to Bob Dylan's use of an electric guitar at the Newport Folk Festival in 1965.

Here is a typical example of the negative reactions to the marketing of cyberpunk in these same musical terms:

> An audience exists, however, for whom Sterling's revolutionary pose, his guerrilla-fighting metaphors, his take-no-prisoners affect, his hipper-than-thou dismissals, his oversimplifications and crude essentialisms, are all appropriate. This audience has little concept of the long historical development of modern SF and furthermore seeks none, but craves merely a new sensation, new refinements of style, and a steady flow of lifestyle accessories with which to reconfirm those refinements, that style. . . . This audience is represented and catered to by editors at *Rolling Stone* and Programmers at MTV. It is for them and their representatives that Sterling presents his tale of SF before and after the heroic arrival of the cyberpunks: lo! Until recently, sci-fi was written by boring nerds, and now it's being invaded by hip young men! . . . Sound all the horns. Sci-fi grows up.
>   It certainly plays better than the complex truth.
>   For fifteen minutes, at any rate. (Patrick Hayden, *MR*, 41)

5. Norman Spinrad (1986) discusses the cyberpunk/New Wave relationship in some detail. He calls *Neuromancer* a "New Wave hard science fiction novel" (p. 183). Benford compares Gibson to Ballard in their concern for surfaces (*MR*, 19). For an overview of Ballard's work, see the special issue of *Re/Search*, 8/9 (1984).

6. Kathy Acker and William Burroughs are notable exceptions: Gibson has been compared to the second and parodied by the first.

7. The most successful such attempts at portraying the near future through the combination of technology and social extrapolation would be John Brunner's three best-known SF novels: *Stand on Zanzibar* (1968—overpopulation), *The Sheep Look Up* (1972—eco-catastrophe), and *The Shockwave Rider* (1975—future shock and the information society).

8. This apparent indifference leads some critics to compare Gibson's tone to a shrug: "Dystopia is already here, say the cyberpunks, and we might as well get used to it" (Hynes 1988, 18).

For the optimistic vision of the future—up to the Second World War—see Joseph J. Corn and Brian Horrigan, *Yesterday's Tomorrows: Past Visions of the American Future* (New York: Summit, 1984); and Joseph J. Corn, ed., *Imagining Tomorrow: History, Technology, and the American Future* (Cambridge: MIT Press, 1986). See also Paul Carter, *The Creation of Tomorrow: Fifty Years of Magazine Science Fiction* (New York: Columbia University Press, 1977). The classic filmic representation of this vision would be H. G. Wells's 1936 collaboration with William Cameron Menzies on the film *Things to Come*.

For the pessimistic and dystopian leanings of postwar science fiction, see Fitting (1979, 1988). Dystopian views of the future are abundant in recent SF, and include such visions of the city of the future as *Escape from New York* (1981) and *Blade Runner*, as well as the much earlier *Soylent Green* (1973). For a recent history of SF, see Brian Aldiss with David Wingrove, *Trillion Year Spree* (London: Paladin, 1988; first published 1986).

9. This leads us to what might be called the Japanese connection, which was already anticipated in that first Pacific Rim novel, Philip K. Dick's 1962 *The Man in the High Castle*, in which Germany and Japan are shown as victors in the Second World War. The novel is set on the West Coast of the United States, which is ruled by the Japanese. This cultural blurring was also evident in the street scenes in *Blade Runner*.

10. There is, at least, no attention drawn to androids in terms of their difference—as opposed to the AIs. The Yakuza assassin of "Johnny Mnemonic" (*BC*), for instance, was "grown in a vat," but there is no discussion of his role in terms of the human/machine dichotomy.

11. The AI's struggle is similar to the plot of *Blade Runner*, which tells of a bounty hunter who "retires" escaped androids. But from Dick's novel *Do Androids Dream of Electric Sheep?* (1968) to the film there had been a change, for, as I have pointed out, the android is a figure of the "nonhuman" for Dick. See my "Futurecop: The Neutralization of Revolt in *Blade Runner*," *Science Fiction Studies*, 14 (1987), 340-54.

12. One of the first and best fictional accounts of a computer-enhanced reality that someone then enters is Vernor Vinge's 1981 story "True Names," *True Names* (New York: Baen, 1987).

13. In Canadian and U.S. law, there has been a trend to recognizing corporations as "persons." For a review of these developments, see H. J. Glasbeek, "The Corporate Social Responsibility Movement—The Latest in Maginot Lines to Save Capitalism," *Dalhousie Law Journal*, 11 (March 1988), 360-80.

14. Total television has long been an SF theme: see, for instance, Ray Bradbury's *Fahrenheit 451*. Another SF variation on the imaginary environments of television is developed in the shared (and drug-induced) hallucinations of the Perky Pat "layouts" (Barbie and her dream house) of Philip K. Dick's 1966 *The Three Stigmata of Palmer Eldritch*.

15. David Porush, *The Soft Machine* (New York: Methuen, 1985), 18.

16. Brian McHale, *Postmodern Fiction* (New York: Methuen, 1987), 59-72.

17. More appropriate in this context would be a contemporary reference to the self-destructing machines of Mark Pauline and the Survival Research Laboratories. For a description of his work, see the "Industrial Culture" issue of *Re/Search*, 6/7 (1983);

also see Jim Pomeroy, "Black Box S-Thetix: Labor, Research, and Survival in the He[Art] of the Beast," in this volume.

18. The definition of SF as serving to acclimatize us to the future was popularized by Alvin Toffler's *Future Shock* (1970). This theme is worked out in John Brunner's novel *Shockwave Rider* (1975). (For a discussion of Toffler and the definition of SF, see Patrick Parrinder, *Science Fiction: Its Criticism and Teaching* [London: Methuen, 1980].) The cyberpunks have rediscovered Toffler (e.g., Sterling 1988, xii).

19. Jameson proposes a possible way of representing the seemingly unrepresentable reality of late capitalism through the concept of "cognitive mapping"—a "situational representation on the part of the individual subject to that vaster and properly unrepresentable totality which is the ensemble of the city's structure as a whole" (FJ, 90)—to the way we "also cognitively map our individual social relationship to local, national and international class realities" (FJ, 91).

> The incapacity to map socially is as crippling to political experience as the analogous incapacity to map spatially is for urban experience. It follows that an aesthetic of cognitive mapping in this sense is an integral part of any socialist political project. (FJ, 353)

REFERENCES

Engelhardt, Tom. "To boldly go where no ad has gone before," *In These Times* (May 17–23, 1989), 18.

Fitting, Peter. "The Modern Anglo-American SF novel: Utopian Longing and Capitalist Cooptation," *Science Fiction Studies,* 6 (March 1979), 59–76.

Fitting, Peter. "Ideological Foreclosure and Utopian Discourse," *Sociocriticism,* 7 (1988), 11–25.

Gibson, William. *Neuromancer* (New York: Ace, 1984).

Gibson, William. *Burning Chrome* (New York: Ace, 1987). (First published 1986.)

Gibson, William. *Count Zero* (New York: Arbor House, 1986).

Gibson, William. *Mona Lisa Overdrive* (Toronto: Bantam, 1988).

Haraway, Donna. "A Manifesto for Cyborgs: Science, Technology, and Socialist Feminism in the 1980s," *Socialist Review,* 80 (1985), 65–107.

Hynes, James. "Robot's rules of disorder: Cyberpunk rocks the boat," *In These Times* (November 23–December 6, 1988), 18–19.

Jameson, Fredric. "Postmodernism, or the Cultural Logic of Late Capitalism," *New Left Review,* 146 (1984), 53–92.

Jameson, Fredric. "Periodizing the 60s," *The 60s without Apology,* ed. Sohnya Sayres et al. (Minneapolis: University of Minnesota Press, 1984).

Jameson, Fredric. "Cognitive Mapping," *Marxism and the Interpretation of Culture,* ed. Cary Nelson and Lawrence Grossberg (Urbana: University of Illinois Press, 1988), 347–60.

Mattelart, Armand, Xavier Delcourt, and Michele Mattelart. *International Image Markets* (London: Comedia, 1984).

Mississippi Review, 47/48 (1988), special issue, "The Cyberpunk Controversy."

Sobchack, Vivian, *Screening Space: The American Science Fiction Film* (New York: Ungar, 1987).

Spinrad, Norman. "On Books: The Neuromantics." *Isaac Asimov's Science Fiction Magazine,* 104 (May 1986), 180–90.

Sterling, Bruce, ed. *Mirrorshades: The Cyberpunk Anthology* (New York: Ace, 1988). (First published 1986.)
Sterling, Bruce. *Islands in the Net* (New York: Ace, 1989).

Contributors

Houston A. Baker, Jr., is a scholar of black literature and culture and a poet who has published three volumes of creative work. He directs the Center for the Study of Black Literature and Culture at the University of Pennsylvania. Recently elected second vice-president of the Modern Language Association of America (with succession to the presidency in 1992), he is currently seeing a book on African-American women writers through its final press production phases. He also holds the Albert M. Greenfield Chair at Pennsylvania and won the Governor's Award for Excellence in the Humanities in 1990.

Sandra Buckley is an associate professor in the Centre for East Asian Studies and a member of the Comparative Literature Program at McGill University in Montreal. She teaches courses in Japanese literature, culture, and film. She has recently completed a volume of interviews and translations entitled *The Broken Silence: Voices of Japanese Feminism* and is also completing a manuscript on Japanese comic books with the working title *Phallic Fantasies: Sexuality and Violence in Japanese Comic Books*. Her research interests focus on the questions and tensions that arise from the application of "Western" critical theory to Japanese cultural texts. She has a particular interest in the comparative study of the practice and theory of feminism.

Peter Fitting is an associate professor of French at the University of Toronto. His articles on utopias and science fiction have appeared in many different journals, including *Cineaction, Science Fiction Studies, Sociocriticism, Utopian Studies,* and *Women's Studies.*

Reebee Garofalo has experience as a performing musician and a benefit concert promoter, and has straddled the worlds of academia and popular music for many years. He has written and lectured widely on topics including the operations of the music business, racism, popular music as an educational tool, and the politics of mega-events. A professor at the University of Massachusetts in Boston, Garofalo is coauthor of *Rock 'n' Roll Is Here to Pay: The History and Politics of the Music Industry.* He is currently doing research for a book on the relationship between popular music and political movements around the world. Garofalo also serves on the U.S. Executive Committee of the International Association for the Study of Popular Music. To preserve his sanity, he enjoys drumming and singing with the Blue Suede Boppers, a fifties rock 'n' roll band.

DeeDee Halleck has been a leading figure in the alternative media movement for the past fifteen years. A founder and producer of Paper Tiger TV and a director of Deep Dish TV, she also teaches communications at the University of California, San Diego.

Donna Haraway is a professor with the History of Consciousness Board, University of California at Santa Cruz. Her teaching and writing focus on feminist theory, women's studies, and science studies with a particular emphasis on the politics, histories, and cultures of modern science and technology. She is the author of *Crystals, Fabrics, and Fields: Metaphors of Organicism in 20th Century Developmental Biology* (1976), *Primate Visions: Gender, Race, and Nature in the World of Modern Science* (1989), and *Simians, Cyborgs, and Women: The Reinvention of Nature* (1990). She is currently writing on the promises of monsters, in and out of science fiction and the fictions of science, for feminist cultural studies.

Valerie Hartouni is assistant professor of communication at the University of California, San Diego. She received her Ph.D. from the History of Consciousness Program at the University of California at Santa Cruz and is currently completing a collection of essays that considers as a problem of discourse and culture the controversies surrounding the development and use of the new technologies of human genetics and reproduction.

Constance Penley is associate professor of English and film studies at the University of Rochester. She is the author of *The Future of an Illusion: Film, Feminism, and Psychoanalysis* (Minnesota, 1989) and editor of *Feminism and Film Theory*. Penley is a founding coeditor of *Camera Obscura: A Journal of Feminism and Film Theory*.

Jim Pomeroy is an associate professor teaching video and photography in the Department of Art and Art History at the University of Texas at Arlington. A general practitioner working in performance, installation, artists' books, and video/audiotape, his art fuses anachronistic technology of the eighteenth and nineteenth centuries with experimental media and new tech. Active in the development of innovative exhibition and performance venues, Pomeroy was one of the founding members of the San Francisco artspace, 80 Langton Street, and has participated in the Exploratorium's artist-in-residence program both as an artist and as an adviser. His previous research includes a 1976 study of the iconography inspired by Mt. Rushmore and an anaglyphic satire of NASA's big science mythology, *Apollo Jest*, published in 1983 as a set of eighty-eight 3D bubblegum cards. In 1988, he co-curated the exhibition "Digital Photography" with Marnie Gillett for San Francisco Camerawork. His most recent performance, the cyber suite IKONIKIRONIK, premiered *Munni Tox*, a computer-animated, digitally sampled rendition of George Bush's inaugural address spoken by all the faces on U.S. currency, from a penny to the hundred-dollar bill.

Processed World is the product of several hundred people's creative efforts over the past decade. These unnamed hundreds have performed almost every job one can think of, from word processor to janitor to hooker to programmer and everything in between. Perhaps the greatest commonality among them is the division imposed by the economy between what they do for money and what they might do if they were free to exercise their own creative inclinations. The magazine continues to publish twice a year and to solicit material from the wage slaves of the world. As broke as ever, *Processed World* depends on people who understand the expression "labor of love," and accept the impossibility of being paid for their honest self-expressions.

Andrew Ross teaches English at Princeton University. He is the author of *Strange Weather: Culture, Science, and Technology in an Age* *of Limits* (1991) and of *No Respect: Intellectuals and Popular Culture* (1989) and the editor of *Universal Abandon? The Politics of Postmodernism* (1988). He is also a member of the Social Text editorial collective.

Paula A. Treichler has been a teacher and administrator at the University of Illinois at Urbana since 1972 and is now associate professor in the College of Medicine's Medical Humanities and Social Sciences Program, the Institute of Communications Research, the Unit for Criticism and Interpretive Theory, and the Women's Studies Program. Her research interests include feminist theory, cultural studies, and the analysis of medical discourse. She is coauthor of *A Feminist Dictionary* (1985) and coeditor of *For Alma Mater: Theory and Practice in Feminist Scholarship* (1985); with Francine Frank, she is coauthor of *Language, Gender, and Professional Writing: Theoretical Approaches and Guidelines for Nonsexist Usage*, published by the Modern Language Association in 1989. She has lectured and published widely on the AIDS epidemic and is completing a book on AIDS and culture.

Index

Compiled by Robin Jackson